Regeneration In Lower Vertebrates And Invertebrates: I.

Papers by
James E. Turner, P. F. A. Maderson,
V. V. Brunst, A. d'A. Bellairs,
Jocelyn M. Zika, David L. Stocum,
Thomas L. Lentz. Wesley P. Norman,
D.J. Procaccini et al.

MSS Information Corporation
655 Madison Avenue, New York, N. Y. 10021

Library of Congress Cataloging in Publication Data
Main entry under title:

Regeneration in lower vertebrates and invertebrates.

  1. Regeneration (Biology). I. Eichler, Victor B.
QH499.R37   592'.03'1   72-8949
ISBN 0-8422-7052-3 (v. 3)

TABLE OF CONTENTS

CREDITS & ACKNOWLEDGEMENTS

Bellairs, A. d'A.; and Susan V. Bryant, "Effects of Amputation of Limbs and Digits of Lacertid Lizards," *Anatomical Record*, 1968, 161:489-496.

Brunst, V. V., "The Effects of Partial Irradiation on Tail Regeneration in Adult *Siredon mexicanum*," *The American Journal of Roentgenology, Radium Therapy and Nuclear Medicine*, 1969, 105:196-206.

Lentz, Thomas L., "Development of the Neuromuscular Junction. I. Cytological and Cytochemical Studies on the Neuromuscular Junction of Differentiating Muscle in the Regenerating Limb of the Newt *Triturus*," *The Journal of Cell Biology*, 1969, 42:431-443.

Maderson, P. F. A.; and P. Licht, "Factors Influencing Rates of Tail Regeneration in the Lizard *Anolis carolinensis*," *Experientia*, 1968, 24:1083-1086.

Maderson, P. F. A.; and S. N. Salthe, "Further Observations on Tail Regeneration in *Anolis carolinensis* (Iguanidae, Lacertilia)," *The Journal of Experimental Zoology*, 1971, 177:185-190.

Norman, Wesley P.; and Anthony J. Schmidt, "The Fine Structure of Tissues in the Amputated-regenerating Limb of the Adult Newt, *Diemictylus viridescens*," *Journal of Morphology*, 1967, 123:271-312.

Procaccini, D. J.; and Catherine M. Doyle, "Histochemical Identification of Sulfated and Non-Sulfated Mucopolysaccharides in Regenerating Forelimbs of Adult Lurodeles," *Oncology*, 1970, 24:313-317.

Stocum, David L., "The Urodele Limb Regeneration Blastema: A Self-organizing System. I. Differentiation *in Vitro*," *Developmental Biology*, 1968, 18:441-456.

Stocum, David L., "The Urodele Limb Regeneration Blastema: A Self-organizing System. II. Morphogenesis and Differentiation of Autografted Whole and Fractional Blastemas," *Developmental Biology*, 1968, 18:457-480.

Turner, James E.; and Samuel R. Tipton, "The Role of the Lizard Thyroid Gland in Tail Regeneration," *Journal of Experimental Zoology*, 1971, 178:63-896.

Zika, Jocelyn M., "A Histological Study of the Regenerative Response in a Lizard, *Anolis carolinensis*," *Journal of Experimental Zoology*, 1969, 172:1-10.

PREFACE

Regeneration of limbs in amphibians and reptiles, regeneration of eye (Woffian regeneration) and regeneration of the basic body plan in Hydra are among the classic systems for the study of differentiation and growth.

Experimental investigation of these systems has proceeded at a high rate within the last five years, as papers in the present collection illustrate. These volumes, confined to the metazoa, consider the fundamental questions of morphogenesis including the possible storage and utilization of undifferentiated cells in the adult form, the means by which cells appear to "know where they are" in developing tissue, the process of dedifferentiation, and the control of regeneration by neural and endocrine factors. Intracellular regeneration, as in the ciliate protozoa, regeneration in plants, and regeneration in planaria including the Turbellarian flatworms are covered in separate volumes.

Tail and Limb Regeneration in Lizards

The Role of the Lizard Thyroid Gland in Tail Regeneration [1,2]

JAMES E. TURNER AND SAMUEL R. TIPTON

Experimental studies on saurian tail regeneration have focused attention primarily on the histology of the normal and regenerate tails as well as the effects of the spinal cord, peripherial nerves and ependyma on the regenerative process (Woodland, '20; Byerly, '25; White, '25; Kamrin and Singer, '55; Jamison, '64; Simpson, '65, '70; Cox, '69a,b). However, little is known about other physiological factors that modify lizard tail regeneration such as age, nutritional requirements, temperature, photoperiod and hormones (Maderson and Licht, '68).

Studies of amphibian regeneration have demonstrated a marked hormonal dependence of regenerative processes, especially in the early stages (Rose, '64; Tassava, '68, '69a,b; Tassava, Chlapowski and Thornton, '68; Liversage and Scadding, '69). Little attention has been given to the potential involvement of endocrines in lizard tail regeneration. To date there

have been only two studies suggesting the dependence of lizard tail regeneration on hormonal control. Both studies used *Anolis* as the experimental animal (Licht and Jones, '67; Licht and Howe, '69). A quantitative relationship was demonstrated between the amount of pars distalis removed and the rate of tail regeneration in *Anolis* (Licht and Howe, '69). In the same study an injection of a combination of adenohypophyseal hormones restored tail regeneration to near normal levels in hypophysectomized lizards. These results definitely demonstrated that the pituitary gland plays an important role in the regulation of tail regeneration in *Anolis*. However, this study failed to spe-

[1] Based on a dissertation presented to the graduate council of the University of Tennessee in partial fulfillment of the requirements for the degree of Doctor of Philosophy.
[2] This research supported by NIH Predoctoral Fellowship 1 F01 GM-42535-01.

cify the degree of importance of any particular one of these adenohypophyseal hormones to the process of lizard tail regeneration. Furthermore, the specific role of other endocrine glands and their secretions in lizard tail regeneration has yet to be clarified. It is with this idea in mind that we propose to demonstrate the role of the thyroid gland in lizard tail regeneration.

MATERIALS AND METHODS
General procedure

The lizards used in this study were obtained from the Louisiana Biological Supply House in Norco, Louisiana. Only adult lizards, ranging from 60 to 70 mm in snout-vent length and 4.0 to 7.5 gm were used in this study. These restrictions were placed in the type of lizards used in order to assure only the utilization of mature Anolis (Fox and Dessauer, '57).

Tails were amputated with scissors 30 mm from their base along the preformed autotomy planes of the epidermis and vertebral column. Preliminary studies have demonstrated that the rate of tail regeneration after such amputation was the same as after autotomy induced by pulling (Licht and Howe, '69). Blastema formation was judged to begin when the first outward signs of growth appeared from around the periphery of the amputation plane, and was judged to be complete when a smooth, black protuberance appeared over the amputation surface. The rate of subsequent tail growth was determined by regular measurements of the growing tissue with a millimeter scale.

Lizards were maintained in specially designed temperature chambers with a regulated photoperiod. All studies were carried out at 32°C. This temperature was chosen for the studies because it was found to be the mean preferred body temperature of Anolis when exposed to a thermal gradient (Licht, '68). Also, 32°C was found to be the most desirable temperature giving optimal results for regenerative tail growth in Anolis (Maderson and Licht, '68).

Photoperiod was provided by a 15-watt bulb in all studies. In all experiments lizards were exposed to a six hour light period per 24 hours. This light period was used because it was found to minimize the effect of photoperiodic stimulation of growth in Anolis (Fox and Dessauer, '57; DiMaggio, '60). All light periods began at 7AM and ended at 1 PM.

Lizards were made hypothyroid by the administration of an oral dose of 0.1 ml of 0.25% thiourea given every other day. Thiourea treatment was begun with all animals at room temperature (21–25°C) six weeks prior to the beginning of the experiments and was continued throughout the course of the studies. It has been previously shown that this dosage of thiourea given over a three week period was sufficient to suppress I^{131} uptake and release in Anolis (Lynn et al., '65). Higher doses of thiourea used with Anolis have been found to be toxic (Ratzersdorf et al., '49; Adams and Craig, '51). Lizards were made hyperthyroid by intraperitoneal injections of 0.2 μg of the sodium salt of L-thyroxine per gm body wt every other day for the entire test period. L-thyroxine was administered in an alkaline solution (10^{-3} N NaOH). Animals were considered hypothyroid when the metabolic rate had fallen 20–30% below the euthyroid level and were considered hyperthyroid when the metabolic rate had risen 20–30% above the euthyroid level (Maher and Levedahl, '59).

The replacement dose of thyroxine given hypothyroid lizards was 0.2 μg per gm body wt every day during the entire test period. All controls received intraperitoneal injections of the alkaline vehicle (10^{-3} N NaOH) given every other day.

Histology

Regenerating tail tips (6 from each group) were taken from hypothyroid lizards as well as from those receiving thyroxine replacement 10, 16 and 24 days after amputation and fixed in Bouin's fixative. Following a 24 hour period of fixation, the tissues were placed overnight in Jenkin's fluid for decalcification and dehydration (Humanson, '62). The tissues were completely dehydrated in absolute alcohol for several hours, cleared in toluene and Cedarwood oil (1:1) overnight and infiltrated with 58°C melting point Tissuemat for six hours. The embedded tissues were sectioned at 10 μ, mounted

11

on slides and stained with Mallory's triple stain (Humanson, '62).

Hypophysectomy

Animals were chilled to 5°C for 15 minutes prior to the operation. An opening into the depression of the basisphenoid bone was achieved by inserting a 26 gauge needle through the soft presphenoid cartilage. The pituitary gland was removed by suction from a blunt ended 23 gauge needle (attached to a 20 cm³ syringe) which had been inserted through the opening in the presphenoid cartilage. The basisphenoid bone which completely surrounded the caudal portion of the pituitary formed a natural guide and barrier for the needle insertion, thus, preventing it from penetrating into the brain tissue.

Histological examination verified the completeness of the operation. The lizard heads were fixed in Bouin's fixative for 24 hours. The tissues were decalcified and dehydrated in Jenkin's fluid (Humanson, '62) overnight, were then completely dehydrated in absolute alcohol for several hours and cleared in a toluene-Cedarwood oil mixture (1:1) overnight. The heads were embedded *in vacuo* in 58°C melting point Tissuemat, sectioned at 15 μ and stained with Mallory's triple stain (Humanson, '62).

After hypophysectomy a period of four days (at room temperature) was allowed before hormone therapy was begun. Three groups of 12 lizards each were given different dosages of thyroxine (0.04 μg, 0.2 μg and 1.0 μg of the sodium salt of L-thyroxine per gram body wt given every other day over the entire test period) and a fourth group served as hypophysectomized controls receiving only the alkaline vehicle (10^{-3} N NaOH). After tail amputation (in a manner previously described) lizards were placed in constant temperature chambers (32°C) with a six hour light period per 24 hours. The time of blastema appearance and the amount of linear tail growth over a three week period were used as criteria for evaluating the various treatments.

Statistical analysis

Analysis of variance was used to find variation within a given set of data. The F test was used to show the significant differences in the variance (Sokal and Rohlf, '69).

RESULTS

Hypothyroidism, hyperthyroidism and thyroxine replacement therapy

Both phases of lizard tail regeneration (blastema formation and tail elongation) appeared responsive to changes in thyroid hormone levels. The completion of blastema appearance was delayed 11 days in the hypothyroid group when compared with controls (fig. 1) and tail elongation was completely inhibited by thiourea treatment (figs. 2, 3). Consequently, the average daily rate of regenerative tail growth (fig. 4) was significantly smaller in the hypothyroid group when compared with controls ($p < 0.001$).

Thyroxine replacement therapy resulted in normal or precocious blastema appearance (fig. 1) and allowed for normal tail elongation up through the fourth week of therapy (figs. 2, 3) after which time the control tail lengths were significantly greater ($p < 0.001$). In addition, the daily rates of regenerative tail growth between controls and thyroxine treated hypothyroid animals were not significantly different until four and five weeks after amputation (fig. 4) at which time the control rates were significantly greater ($p < 0.001$).

Hyperthyroidism caused an early appearance of blastema emergence (fig. 1). However, figure 2 shows that there was no significant difference between tail lengths during the entire test period when controls were compared with hyperthyroid animals ($p > 0.05$). However, there was a slight precocious increase and subsequently an early tapering off of the hyperthyroid regenerate growth rate when compared with controls (fig. 4). Two weeks after amputation the rate of regenerative tail growth was significantly greater in the hyperthyroid group when compared with controls ($p < 0.05$). However, the hyperthyroid growth rate was not significantly different from those of controls at the end of three and four weeks ($p > 0.05$), but was significantly less than that of controls at the end of five weeks ($p < 0.001$). In addition, the

12

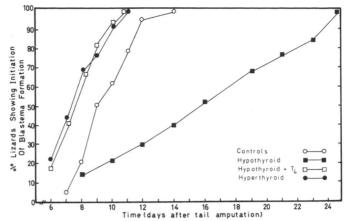

Fig. 1 The effect of various thyroid states on initiation of blastema formation in *Anolis* at 32°C. Values represent percent lizards within each group showing initial signs of blastema formation. T_4, 0.2 μg/gm body wt L-thyroxine given every other day. Each group contains 15–20 individuals.

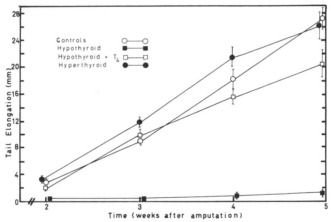

Fig. 2 The effect of various thyroid states on tail elongation in *Anolis* at 32°C. Values represent mean length and vertical lines indicate ± one standard error. For T_4 value see figure 1. Each group contains 12–15 individuals.

average daily rate of regenerative tail growth achieved by both controls (0.90 mm/day) and hyperthyroid lizards (0.92 mm/day) for the entire test period (weeks 1–5) was the same.

Hypophysectomy

Hypophysectomy was found to delay blastema formation by one and one-half weeks (fig. 5) and to inhibit tail elonga-tion in the same manner as was found in hypothyroid lizards (figs. 6, 7). Hypo physectomized lizards showed no appre-ciable tail growth two weeks after ampu-tation as compared with 2.3 ± 0.3 mm reported for intact controls. At the end of three weeks hypophysectomized lizards had obtained a mean tail length of 2.3 ± 0.6 mm in contrast to a length of 9.5 ± 0.6 mm reported for intact controls. Con-

Controls

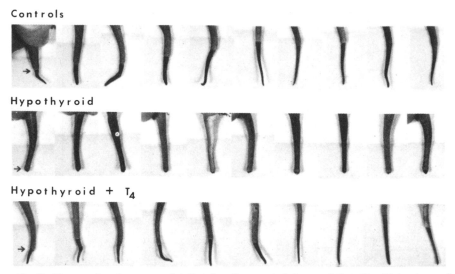

Hypothyroid

Hypothyroid + T$_4$

Fig. 3   Photographs of regenerated tails taken from control (top row), hypothyroid (center row), and hypothyroid with thyroxine treatment (bottom row) *Anolis* 28 days after amputation (see fig. 2 for quantitative analysis). Arrows indicate site of amputation where new regeneration begins. For T$_4$ value see figure 1. The ten regenerates in each row are a representative number from groups containing 15–20 individuals.

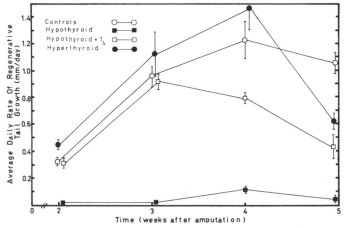

Fig. 4   The effect of various thyroid states on the average daily rate of regenerative tail growth during weekly intervals in *Anolis* at 32°C. Values represent mean length and vertical lines indicate ± one standard error. For T$_4$ value see figure 1. Each group contains 12–15 individuals.

sequently, two and three weeks after amputation tail lengths of hypophysectomized lizards were significantly less than those of intact controls ($p < 0.001$).

The results from thyroxine replacement therapy given every other day over a three week period to hypophysectomized lizards are indicated in figures 5, 6 and 7.

14

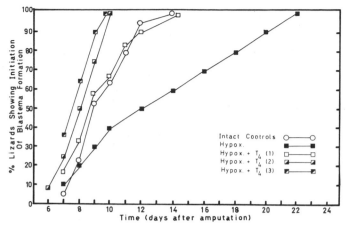

Fig. 5 The effect of hypophysectomy and thyroxine replacement therapy on the initiation of blastema formation in *Anolis* at 32°C. Values represent percent lizards within each group showing initial signs of blastema formation. T_4 (1), 0.04 μg/gm body wt L-thyroxine given every other day; T_4 (2), 0.2 μg; T_4 (3), 1.0 μg. Each group contains 12–15 individuals.

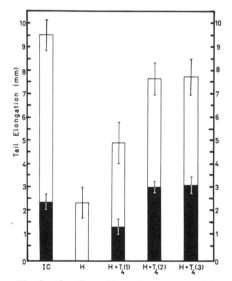

Fig. 6 The effect of hypophysectomy and thyroxine replacement therapy on tail elongation in *Anolis* at 32°C. Vertical bars indicate mean length; shaded areas of bars represent tail growth two weeks after amputation; unshaded areas represent tail growth three weeks after amputation. Vertical lines indicate ± one standard error. IC, intact controls; H, hypophysectomized lizards. For T_4 values see figure 5. Each group contains 12–15 individuals.

A dose of 0.04 μg/gm body wt was sufficient to allow for blastema formation closely approximating that of intact controls. The mean tail lengths achieved at this dose level were 1.7 ± 0.3 mm at the end of two weeks and 4.8 ± 0.8 mm at the end of three weeks. This growth was not sufficient to sustain subsequent tail elongation at a rate equivalent to that just reported for intact controls, but was nevertheless significantly greater than growth reported for hypophysectomized lizards ($p < 0.001$).

A dose of 0.2 μg/gm body wt initiated the early development and subsequently the early completion of blastema formation in the same manner as did hyperthyroidism. The mean tail lengths achieved at this dose level were 2.9 ± 0.2 mm at the end of two weeks and 7.6 ± 0.6 mm at the end of three weeks. Therefore, this dose level was sufficient to sustain tail elongation at a rate significantly greater ($p < 0.05$) than the lowest thyroxine dose (0.04 μg) but was still significantly less than that of intact controls ($p < 0.01$). In addition, a dose of 1.0 μg/gm body wt produced the same effect on blastema formation as did the 0.2 μg dose (fig. 5). Also, there was no significant difference between the regenerating tail lengths of those hypophysectomized

15

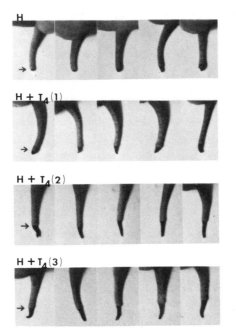

Fig. 7 Photographs of regenerated tails taken from hypophysectomized (top row), and hypophysectomized with various thyroxine treatments (bottom 3 rows) *Anolis* 21 days after amputation (see fig. 6 for quantitative analysis). Arrows indicate site of amputation where new regeneration begins. For T_4 values see figure 5. The five regenerates in each row are a representative number from groups containing 12–15 individuals. H, hypophysectomized lizards.

lizards receiving doses of either 0.2 μg or 1.0 μg of thyroxine (p > 0.05).

These results indicate that a dose response relationship may exist between certain levels of thyroxine (0.04 to 0.2 μg) and tail regeneration in hypophysectomized *Anolis*.

Histology of hypothyroid and thyroxine replacement regenerates

Ten days after amputation the outward growth of the ependymal vesicle past the amputation pland and into the surrounding blastema had already taken place in the thyroxine treated hypothyroid lizards (figs. 8, 10). However, figures 9 and 11 show that in ten day hypothyroid regenerates the spinal cord (from where the ependymal outgrowth occurs) was pinched off and had retracted from the amputation plane. Also, a fibrous connective tissue pad partially covered the amputation surface of the severed spinal cord. In addition a large portion of the hypothyroid regenerate was still covered by the scab that originally formed over the amputation surface. In contrast, only a small portion of the original scab remained on the amputation surface of the thyroxine treated hypothyroid blastema (fig. 8). This evidence, plus the fact that the amount of blastema tissue present in the regenerating tail tips was less in the hypothyroid lizards than in those receiving thyroxine replacement, indicated that this phase of tail regeneration (blastema formation) was retarded in the hypothyroid condition.

Figures 12 and 14 show that 16 days after amputation the ependymal tube in the thyroxine treated hypothyroid regenerates extended the entire length of the elongating tail. Also, in the same photomicrographs can be seen the appearance of procartilage and promuscle aggregates. However, in the 16 day hypothyroid regenerates the spinal cord still remained pinched off and had retracted even further from the amputation plane than in the ten day hypothyroid regenerates (figs. 13, 15). In addition, promuscle aggregates were present as in the 16 day thyroxine treated regenerates but no procartilage aggregate had formed at that time.

At 24 days after amputation tissue differentiation had taken place in the thyroxine treated hypothyroid regenerates and was quite complete particularly at the base of the regenerating tails (figs. 16, 18). Figure 16 shows the complete differentiation of the cartilage tube, several muscle groups and scales. Also shown in figure 16 and in more detail in figure 18 was the extension of the ependymal tube from below the amputation plane into the cartilage tube. Figure 20 shows that the cartilage tube and the enclosed ependyma extended to the tip of the 24 day thyroxine treated hypothyroid regenerates and figure 21 is a higher magnification of the process of scale formation previously seen in figure 16.

In contrast, figure 17 shows the 24 day hypothyroid regenerate. Note that the formation of the ependymal vesicle was

16

delayed by about two weeks when compared with the thyroxine treated regenerates. Furthermore, the ependyma had not as yet grown into the newly differentiated hypothyroid blastema. Figure 19 is a higher magnification showing in more detail the late formed ependymal vesicle. In addition, figure 17 demonstrates the presence of mature muscle fibers as well as tufts of cartilage found at the tips of the severed vertebrae. Figures 22–25 are further photomicrographs of figure 17 whose sections were cut somewhat lateral to those in the above figure and they further demonstrate the process of tissue differentiation within the 24 day hypothyroid blastema. In particular figure 22 shows the formation of four new muscle groups as well as beginning scale formation as was found in the 24 day thyroxine treated regenerates (see figs. 16, 20, 21). In addition, as reported above, tufts of cartilage were formed at the tips of the severed vertebrae. Figure 23 is a higher magnification of figure 22 showing in more detail scale formation at the tip of the blastema by epidermal invagination. The process of scale formation at the tip of a regenerating tail signifies cessation of tail elongation (Cox, '69a). Thus, it can be assumed that no further appreciable tail elongation occurred after this time period in hypothyroid lizards. Figure 23 is a higher magnification of the two most distal of the four newly differentiated muscle groups seen in figure 22. The two most proximal of the four muscle groups are seen in more detail in figure 25.

DISCUSSION

Hypothyroidism in *Anolis* was found to delay blastema formation and to completely inhibit the tail elongation phase of regeneration. This suggests that the first phase of lizard tail regeneration is not as dependent on thyroxine as the tail elongation phase. Similar results were obtained by Ghidoni ('48) when he administered thiourea to the tadpoles of *R. catesbeiana* and *R. pipiens*. Thiourea treatment that commenced prior to or with transection of the tadpole tail and continued until the sixth day after amputation appeared to have little effect on the initial phases of regeneration. Regeneration was retarded only after the sixth day. This suggests that a critical phase in regeneration required the presence of thyroid secretions and the cell proliferation phase (tail elongation) is a notable feature of the sixth day in the tadpole tail.

In the present study thyroxine replacement therapy in hypothyroid lizards allowed for normal blastema formation as well as subsequent tail elongation. Similar results have been reported for urodeles. Thyroxine replacement therapy given to thyroidectomized newts with amputated limbs restored their regenerative capacity (Richardson, '40, '45; Schmidt, '58, '68). In the present study a dose of only 0.2 μg/gm body wt of L-thyroxine given every other day over a five week period restored the regenerative capacity in hypothyroid *Anolis*. Schmidt ('58) also found that regeneration resulting from low titers of thyroxine (0.14 μg) diffusing from pellets implanted in the amputated forelimb of thyroidectomized newts was not different from that of euthyroid newts.

Hyperthyroidism had no effect on the tail elongation phase in *Anolis*; however, there was evidence of precocious blastema development. It had been demonstrated earlier that large doses of thyroid hormone given to either hypo- or euthyroid newts elicited a typical hyperthyroid inhibition (Kambara, '53; Schmidt, '58); however, this was not the case in the present study.

The effect of hypophysectomy on *Anolis* tail regeneration reported here was the same as that described by Licht and Howe ('69). Hypophysectomy caused a delay in blastema formation by one and one-half weeks and completely inhibited tail elongation. Here again, as with the hypothyroid lizards, it appears as if the blastema phase of tail regeneration was not as dependent on hormonal control as was the tail elongation phase. It is entirely possible that the lizard tail blastema can develop and differentiate as a self-organizing system in much the same manner as was reported for the urodele regeneration blastema (Stocum, '68a,b).

A dose-response relationship was found to exist over a certain range of thyroxine concentrations (0.04 to 0.2 μg) for tail regeneration in hypophysectomized *Anolis*. The intermediate dose of thyroxine (0.2 μg) produced the maximum amount of tail elongation which was still significantly less than that of intact controls. A larger dose of thyroxine (1.0 μg) did not further augment the regeneration process and it was postulated that a dose much above this level would elicit an inhibitory effect. Similarly, Schmidt ('58) has shown that a dose-response relationship exists between certain thyroxine levels and limb regeneration in newts. In addition, he reported that retardation of regeneration was the rule when more than a certain concentration of this hormone was used. It therefore appears that there is a certain dose of thyroxine which produces the maximum thyroid-tail regeneration response hypophysectomized as well as hypothyroid *Anolis*. This level of thyroxine appears to be approximately 0.2 μg/gm body wt given every other day during the entire test period. In summary, it could be said that normal tail regeneration in *Anolis* is dependent in part upon adequate circulating levels of thyroxine. However, the optimum dose of thyroxine given to hypophysectomized lizards are not adequate enough to elicit a regeneration response equivalent to that of intact controls, although this same level (0.2 μg) given to hypothyroid lizards can bring the regeneration response close to that of intact controls. This fact in turn leads to the conclusion that in addition to thyroxine, lizard tail regeneration is also dependent to some extent on the presence of other hormones to elicit a complete regeneration response. It is possible that prolactin acting synergistically with thyroxine in hypophysectomized lizards may be necessary in order to bring tail growth to within the range of intact controls. In amphibian limb regeneration it was found that a combination of prolactin and thyroxine, but neither alone, was able to significantly improve survival of hypophysectomized adult newts and also maintain the normal ability to regenerate amputated limbs (Tassava, '69a). Gona et al. ('70) have shown that thyroxine acts in either a synergistic or antagonistic way with prolactin in the newt depending on the dosage. This type of mechanism might be a consideration in the growth stimulation-inhibition dosage effects of thyroxine discussed above, particularly since Licht and Jones ('67) have shown that growth

in lizard tail regenerates is influenced by prolactin.

Histological studies revealed that although hypothyroidism retarded blastema formation and inhibited tail elongation, the tissues within the blastema did differentiate along the same time course as those found in hormonal replacement regenerates. An explanation of this phenomenon is proposed as follows. The delayed blastema formation phase in hypothyroid lizards was completed at the same time that the thyroxine treated and control regenerates were about to begin the cell differentiation phase. At this time hypothyroid lizards were found to skip the tail elongation phase and proceed directly into the cell differentiation phase along with the thyroxine treated and control lizards. Consequently, these results indicate that thyroxine had little to do with the tissue differentiation phase of *Anolis* tail regeneration.

An attempt was made to determine the nature of the hypothyroid inhibition of tail elongation in *Anolis*. Histological studies revealed that the principle tissue affected by hypothyroidism was the spinal cord and its ependymal epithelium lining. In hypothyroid lizards there was early evidence of spinal cord regression at the site of the developing blastema which resulted in retardation of the formation of the ependymal vesicle by two weeks and in turn was followed by the inhibition of the outgrowth of the vesicle into the developing blastema. Thyroxine replacement insured the normal development and growth of the ependyma from the spinal cord into the blastema. Kamrin and Singer ('55) demonstrated the dependance of *Anolis* tail regeneration on the presence of the spinal cord. More specifically they showed that destruction of the spinal cord proximal to the autotomy plane resulted in a failure of the tail to regenerate. Later Simpson ('64, '65, '70) modified this view by conclusively demonstrating that the initiation of tail regeneration in the lizard *Lygosoma laterale* was dependent upon the presence of the ependymal epithelium lining the spinal cord and not dependent on the presence of a certain threshold number of central or peripheral nerve fibers as was reported for amphibian limb regeneration by Singer ('47, '52,

'63, '65). Furthermore, Simpson (unpublished) has also shown that the ependymal implants can induce tail regeneration in *Anolis* which had several segments of the spinal cord removed just anterior to the amputation plane. Cox ('69b) further confirmed Simpson's unpublished data for *Anolis* and concluded again that the outgrowth of the ependyma into the developing blastema supported tail regeneration and that the peripheral nerve supply to the area made no significant contribution to the regeneration process.

The histological and morphological observations reported in the present study for hypothyroid regenerates were very similar to those reported by Cox ('69b) who prevented ependymal outgrowth in *Anolis* by spinal cord removal. Cox reported that at 27°C and with the spinal cord removed, the amputation wounds healed in 12 to 14 days and a few days later began forming darkly pigmented mounds approximately 0.5 mm in length. The present study carried out at 32°C demonstrated that wounds healed somewhat earlier and by days 8 to 10 began forming darkly pigmented mounds (blastemas) approximately 0.5 mm long. Cox also reported that the newly formed blastemas in animals in which the spinal cord had been removed, looked much like normal regenerates; however, they grew only 0.7 to 0.8 mm and after two weeks began to form scales. Similarly, growth data from the present study indicated that at the end of the five week test period hypothyroid regenerates attained a length of only 1.0 mm and scale formation began to appear two to three weeks after amputation. In addition, Cox reported that early aggregations of promuscle cells were formed in the small blastemas like those found in the normal regenerates and along the same time course. In the present study the promuscle aggregates were also reported to form early in the hypothyroid regenerates (day 16) as was also the case with the thyroxine treated regenerates. In those lizards in which the spinal cord had been removed, Cox reported that the promuscle aggregates completed differentiation (day 35) giving rise to three or four segments of regenerated muscle just as scale formation began to appear at the tips of the small

blastemas. Similarly, in our study as scale formation began to take place at the tips of the hypothyroid blastemas (day 24), the promuscle aggregates also completed differentiation giving rise to four segments of regenerated muscle.

Cox reported that there was no cartilage formation and consequently no cartilage tube formation in the regenerating tails in which the spinal cord had been removed. In our study no cartilage tube was formed in hypothyroid regenerates as well; however, by day 24 small tufts of cartilage were found at the tips of the severed vertebrae. The reason for the presence of cartilage tufts found in our study and the absence of cartilage in Cox's work can be explained on the basis of the different states of the ependyma and spinal cord in the two studies. In Cox's work the ependyma along with the spinal cord was obliterated and never reformed; however, in the hypothyroid regenerates the ependyma appeared in the form of the vesicle two weeks later than in thyroxine treated hypothyroid lizards but never grew past the amputation plane into the developing blastema. Simpson ('64, '65, '69) demonstrated that one of the functions of the ependyma other than as the initiator of lizard tail regeneration was the induction of cartilage formation. He reported that the amount of cartilage tube that differentiates in a given regenerate is roughly proportional to the amount of ependyma present in the regenerate. Therefore, small tufts of cartilage at the tips of the severed vertebrae in the present study should indicate the presence of a small amount of ependymal tissue. Histological data have shown this to be the case.

It was concluded that hypothyroidism inhibited tail elongation in *Anolis* by a retardation of the formation of the ependymal vesicle and the prevention of its subsequent growth into the developing blastema. In contrast, thyroxine treatment in hypothyroid regenerates was shown to reestablish normal blastema development and subsequent tail elongation in much the same manner as was reported for normal regenerates by Cox ('69a). It has also been shown that thyroxine stimulates ependymal proliferation in the brain of developing amphibians

(Weiss and Rossetti, '51; Pesetsky, '69a). Histochemical investigations have shown markedly increased activity of nucleoside phosphatases as well as enhanced oxidative enzyme activity in the ependymoglia of thyroxine-stimulated developing tadpole brains (Pesetsky, '65, '69b). Also, Pesetsky ('69c) has shown through electron microscopic studies of the brains of thyroprivic *Rana pipiens* larvae that ependymal cells lining ventricles of the developing brain displayed sparsely distributed ribosomes, few polyribosomes, and relatively simple profiles of rough-surfaced endoplasmic reticulum. However, when thyroxine was administered to such animals, free ribosomes and polyribosomes became numerous and stacks of lamellar ribosome-studded endoplasmic reticulum were predominant. It is entirely possible that the observed effect of hypothyroidism and thyroxine replacement therapy on the lizard ependyma represents the same type of phenomenon reported above for amphibians.

It seems possible that the trophic effect that the ependyma has on the developing lizard tail may be the same as that of the action of nerve fibers on anuran and urodele limb regeneration. Singer ('52, '65) has postulated a trophic substance produced by the nerve cell body and transmitted by the axon. That such a substance may be able to pass directly from the nerve cell body into the surrounding tissues was suggested by the experiments of Kamrin and Singer ('59) in which study implanted ganglia were able to support limb regeneration in the newt, *Triturus*. The ependyma is of the same origin as are the nerve cells and is believed to give rise to new nerve cells in a regenerating salamander tail (Duesburg, '25). As has been suggested by Simpson ('64), it does not seem unreasonable to assume that the ependyma may have the same mediating action as that hypothesized for nerve fibers.

All factors taken into consideration, it seems possible that this regenerating system will prove to be an ideal one for further study of thyroid-neuroglial relationships and even a more intriguing one for study considering the aspect of the trophic function assigned to the ependymal cells of this system.

LITERATURE CITED

Adams, A. E., and M. Craig 1951 The effects of antithyroid compounds on the adult lizard *Anolis carolinensis.* J. Exp. Zool., *117:* 387–315.

Byerly, T. C. 1925 Note on the partial regeneration of the caudal region of *Sphenodon punctatum.* Anat. Rec., *30:* 61–66.

Cox, P. G. 1969a Some aspects of tail regeneration in the lizard, *Anolis carolinensis.* I. A description based on histology and autoradiography. J. Exp. Zool., *171:* 127–150.

—— 1969b Some aspects of tail regeneration in the lizard, *Anolis carolinensis.* II. The role of the peripheral nerves. J. Exp. Zool., *171:* 151–160.

DiMaggio, A., III 1960 Hormonal replacement of photoperiod as a stimulus for growth in a lizard. Fed. Proc., *19:* 52 (Abstract).

Duesburg, J. 1925 La regeneration des tissues dans la queue des urodeles. Cellule, *35:* 29–46.

Fox, W., and H. C. Dessauer 1957 Photoperiodic stimulation of appetite and growth in the male lizard, *Anolis carolinensis.* J. Exp. Zool., *134:* 557–575.

Ghidoni, M. 1948 Effects of thyroid inhibitors upon tail regeneration in the tadpole. Growth, *12:* 181–202.

Gona, A. G., T. Pearlman and W. Eakin 1970 Prolactin-thyroid interaction in the newt, *Diemictylus viridescens.* J. Endocrinol., *48:* 585–590.

Humanson, G. L. 1962 Animal Tissue Techniques. W. H. Freeman and Co., San Francisco.

Jamison, J. P. 1964 Regeneration subsequent to intervertebral amputation in lizards. Herpetologica, *20:* 145–149.

Kambara, S. 1953 The effect of thyroid and thiourea on regeneration of the tail of the newt, *Triturus pyrrhogaster.* Annot. Zool. Jap., *26:* 208–212.

Kamrin, R. P., and M. Singer 1955 The influence of the spinal cord in regeneration of the tail of the lizard *Anolis carolinensis.* J. Exp. Zool., *128:* 611–627.

Licht, P. 1968 Responses of the thermal preferendum and heat resistance to thermal acclimation under different photoperiods in the lizard *Anolis carolinensis.* Am. Midland Nat., *79:* 149–158.

Licht, P., and N. R. Howe 1969 Hormonal dependence of tail regeneration in the lizard *Anolis carolinensis.* J. Exp. Zool., *171:* 75–83.

Licht, P., and R. E. Jones 1967 Effects of exogenous prolactin on reproduction and growth in adult males of the lizard *Anolis carolinensis.* Gen. Comp. Endocrinol., *12:* 526–535.

Liversage, R. A., and S. R. Scadding 1969 Reestablishment of forelimb regeneration in adult hypophysectomized *Diemictylus viridescens* given frog anterior pituitary extract. J. Exp. Zool., *170:* 381–396.

Lynn, W. G., J. J. McCormick and J. C. Gregorek 1965 Environmental temperature and thyroid function in the lizard, *Anolis carolinensis.* Gen. Comp. Endocrinol., *5:* 587–595.

Maderson, P., and P. Licht 1968 Factors influencing rates of tail regeneration in the lizard *Anolis carolinensis.* Experientia, *24:* 1083–1086.

Maher, M. J., and B. H. Levedahl 1959 The effect of the thyroid gland on the oxidative metabolism of the lizard, *Anolis carolinensis.* J. Exp. Zool., *140:* 169–189.

Pesetsky, I. 1965 Thyroxine-stimulated oxidative enzyme activity associated with precocious brain maturation in anurans. Gen. Comp. Endocrinol., 5: 411–417.

———— 1969a Autoradiographic analysis of thyroxine-stimulated ependymal cell proliferation and migration in *Rana pipiens.* Am. Zoologist, 9: 1123 (Abstract).

———— 1969b Altered thyroid state and changes in nucleoside phosphatase activity in ependymoglial cells of anuran amphibians. Gen. Comp. Endocrinol., Suppl., 2: 238–244.

Pesetsky, I., and P. G. Model 1969 Thyroxine-stimulated ultrastructural changes in ependymoglia of thyroprivic amphibian larvae. Exp. Neurol., 25: 238–245.

Ratzersdorfer, C., A. S. Gordon and H. A. Charipper 1949 The effects of theourea on the thyroid gland and the molting behavior of the lizard, *Anolis carolinensis.* J. Exp. Zool., *112:* 13–27.

Richardson, D. 1940 Thyroid and pituitary hormones in relation to regeneration. I. The effect of anterior pituitary hormones on regeneration in hind leg in normal and thyroidectomized newts. J. Exp. Zool., 83: 407–425.

———— 1945 Thyroid and pituitary hormones in relation to regeneration. II. Regeneration of the hind leg of the newt, *Triturus viridescens,* with different combinations of thyroid and pituitary. J. Exp. Zool., 100: 417–427.

Rose, S. M. 1964 Regeneration. In: Physiology of the Amphibia. J. A. Moore, ed. Academic Press, New York, pp. 345–622.

Schmidt, A. J. 1958 Thyroxine effects on forelimb regeneration in the adult newt, *Triturus viridescens.* Anat. Rec., *132:* 503–504.

———— 1968 Endocrine secretions. In: Cellular Biology of Vertebrate Regeneration and Repair. Un. Chicago Press, Chicago, pp. 245–289.

Simpson, S. B., Jr. 1964 Analysis of tail regeneration in the lizard *Lygosoma laterale.* I. Initiation of regeneration and cartilage differentiation: the role of the ependyma. J. Morph., *114:* 425–435.

———— 1965 Regeneration of the lizard tail. In: Regeneration in Animals and Related Problems. V. Kiortsis and H. A. L. Trampusch, eds. North-Holland Pub. Co., Amsterdam, pp. 431–443.

———— 1969 Studies on lizard tail regeneration. Am. Zool., 9: 597 (Abstract).

———— 1970 Studies on regeneration of the lizards tail. Am. Zool., 10: 157–165.

Singer, M. 1947 The nervous system and regeneration of the forelimb of adult *Triturus.* VII. The relation between nerve fibers and surface area of amputation. J. Exp. Zool., 104: 251–265.

———— 1952 The influence of the nerve in regeneration of the amphibian extremity. Rev. Biol., 27: 169–200.

———— 1965 A theory of the trophic nervous control of amphibian limb regeneration, including a re-evaluation of quantitative nerve requirements. In: Regeneration in Animals and Related Problems. V. Kiortsis and H. A. L. Trampusch, eds. North-Holland Pub. Co., Amsterdam, pp. 20–32.

Singer, M., and E. Mutterperl 1963 Nerve fiber requirements for regeneration in forelimbs of the newt, *Triturus.* Develop. Biol., 7: 180–191.

Sokal, R. F., and F. J. Rohlf 1969 Biometry: The Principles and Practices of Statistics in Biological Research. W. H. Freeman, San Francisco.

Stocum, D. L. 1968a The urodele limb regeneration blastema. A self organizing system: I. Differentiation *in vitro.* Develop. Biol., *18:* 441–456.

———— 1968b The urodele limb regeneration blastema. II. Morphogenesis and differentiation of qutographed and fractional blastemas. Develop. Biol., *18:* 457–480.

Tassava, R. A. 1968 Limb regeneration of hypophysectomized adult newts. Am. Zoologist, 8: 785–786.

———— 1969a Hormonal and nutritional requirements for limb regeneration and survival of adult newts. J. Exp. Zool., *170:* 33–54.

———— 1969b Survival and limb regeneration of hypophysectomized newts with pituitary xenografts from larval axolotls, *Ambystoma mexicanum.* J. Exp. Zool., *171:* 451–457.

Stocum, D. L., F. J. Chlapowski and C. S. Thornton 1968 Limb regeneration in *Ambystoma* larvae during and after treatment with adult pituitary hormones. J. Exp. Zool., *167:* 157–163.

Umbreit, W. W., R. H. Burris and J. F. Stauffer 1957 Manometric Techniques. Burgess Pub. Co., Minneapolis.

Weiss, P., and F. Rossetti 1951 Growth responses of opposite sign among different neuron types exposed to thyroid hormone. Proc. Natl. Acad. Sci. U. S., 37: 540–556.

White, C. P. 1925 Regeneration of the lizards tail. J. Path. Bact., 28: 63–68.

Woodland, W. N. F. 1920 Some observations on caudal anatomy and regeneration on gecko (*Hemidactylus flaviviridis,* Puppel), with notes on tails of *Sphenodon* and *Pygopus.* Quart. J. Micro. Sci., 65: 63–100.

21

PLATES

Abbreviations

a, adipose tissue   m, muscle
b, blastema   nf, nerve fibers
c, cartilage   p, pigmented area
ct, connective tissue   pc, procartilage aggregates
ctu, cartilage tube   pm, promuscle
e, ependyma   s, scab
es, ependymal sac   sc, spinal cord
et, ependymal tube   sca, scales
ev, ependymal vesicle   v, vertebrae
fct, fibrous conn. tissue we, wound epithelium

PLATE 1

EXPLANATION OF FIGURES

8 Tip of ten day hypothyroid regenerate given thyroxine. Horizontal section through mid-dorsal region. Note the outward growth of the ependymal vesicle (ev) past the amputation plane (arrow) into the blastema (b). × 12.7.

9 Hypothyroid regenerate ten days after amputation. Horizontal section. Note that the spinal cord (sc) is pinched off and has retracted from the amputation level (arrows). Also, note the presence of a fibrous connective tissue pad (fct) which is shown partially covering the amputation level (arrows) of the severed spinal cord (sc). × 12.7.

10 Higher magnification of figure 8 showing the ependymal vesicle (ev) and ependymal sac (es) in more detail. × 32.

11 Higher magnification of figure 9 showing the tip of the pinched off and retracted spinal cord (sc). × 32.

PLATE 1

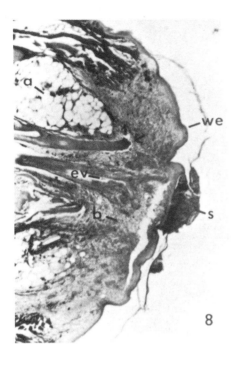

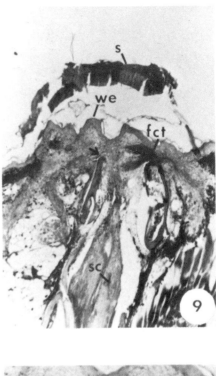

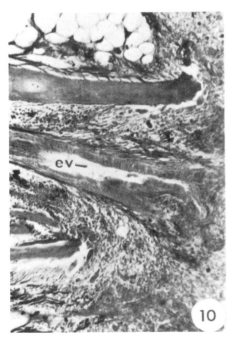

12 Tip of 16 day hypothyroid regenerate given thyroxine. Horizontal section. Regenerate 4 mm in length. Note that the ependymal tube (et) now extends the entire length of the regenerate and at its tip is an area of increased vascularization (arrow). Also, note the appearance of the promuscle (pm) and procartilage (pc) aggregates. × 12.7.

13 Hypothyroid regenerate 16 days after amputation. Horizontal section. Regenerate 1.5 mm in length. Note that the spinal cord (sc) still remains pinched off and has retracted even further from the amputation plane (arrows) than in the ten day hypothyroid regenerate (fig. 9). Also, there is an area of high vascularization (arrow) at the tip of the blastema as in figure 12 but it is not associated with the ependyma. In addition note the appearance of the promuscle aggregates (pm) as in figure 12 but no procartilage aggregate has formed at this time. × 12.7.

14 Higher magnification of figure 12 showing the promuscle (pm) and procartilage (pc) aggregates as well as the ependyma (e) and ependymal tube (et) in more detail. × 32.

15 Higher magnification of figure 13 showing the tip of the pinched off and retracted spinal cord (sc). × 32.

26

PLATE 2

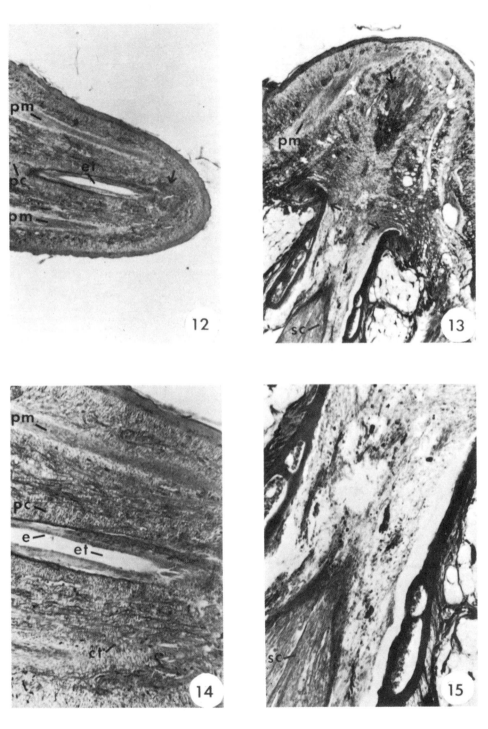

16   Base of the 24 day hypothyroid regenerate given thyroxine. Horizontal section. Regenerate approximately 12 mm in length. Note the complete differentiation of the cartilage tube (ctu), several muscle groups (m) and scales (sca) at the base of this regenerate. Also, note the extention of the ependymal tube (et) from below the amputation plane (arrows) into the cartilage tube. × 12.7.

17   Hypothyroid regenerate 24 days after amputation. Horizontal section. Regenerate 2 mm in length. Note the delayed formation of the ependymal vesicle (ev) which has not as yet extended into the newly differentiated blastema (b). Also, note the presence of mature muscle fibers (m) as seen in figure 16 as well as tufts of cartilage (c) at the tips of the severed vertebrae (v) which marks the amputation plane (arrows). × 12.7.

18   Higher magnification of figure 16 showing the base of the cartilage tube (ctu) as well as the ependymal tube (et) with its associated nerve fibers (nf). Arrow marks amputation plane. × 32.

19   Higher magnification of figure 17 showing the late formed ependymal vesicle (ev). × 32.

28

PLATE 3

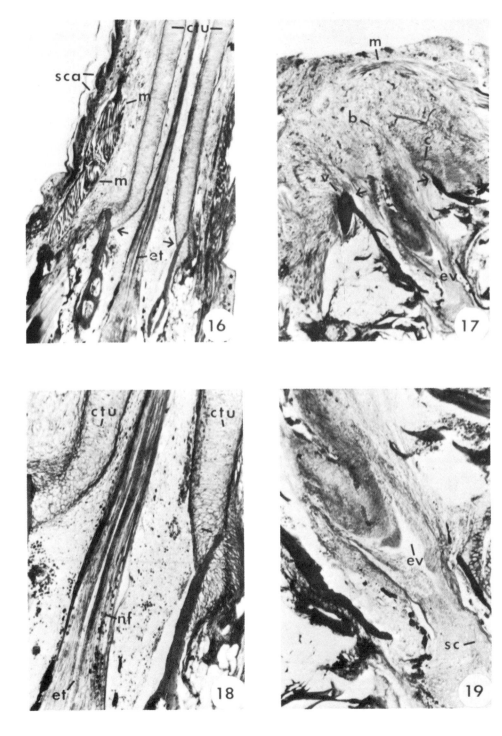

PLATE 4

20 Further photomicrographs of figure 16. Horizontal section. Note the radiating nerve fibers (nf) passing through the opening at the tip of the cartilage tube (arrow). × 32.

21 Further photomicrographs of figure 16. Horizontal section. View of scale formation (sca) by epidermal invagination. In addition, note the heavily pigmented area (p) which gives to the regenerate its characteristic black color. Also, note the proximal portion of a newly differentiated muscle group (m) as well as a portion of the cartilage tube (ctu). × 32.

PLATE 4

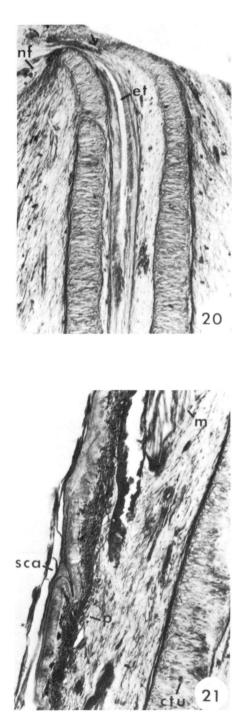

22 Further photomicrographs of figure 17. Horizontal section cut more lateral to that in figure 17. Differentiating hypothyroid blastema (b). Note the formation of four new muscle groups (m) as well as beginning scale formation (sca) like that found in the 24 day thyroxine treated regenerate (fig. 16). In addition, tufts of cartilage (c) are formed at the tips of the severed vertebrae (v) which marks the original amputation plane (arrows). Also, note that scale formation (sca) has begun at the tip of the regenerate. × 12.7.

23 Higher magnification of figure 22 showing scale formation (sca) by epidermal invagination in more detail. In addition, the two most distal of the four newly differentiated muscle groups (m) can be seen. × 32.

24 Higher magnification of figure 22 showing the cartilage tufts (c) in more detail. × 32.

25 Higher magnification of figure 22 showing the two most proximal of the four newly differentiated muscle groups (m) in more detail. × 32.

PLATE 5

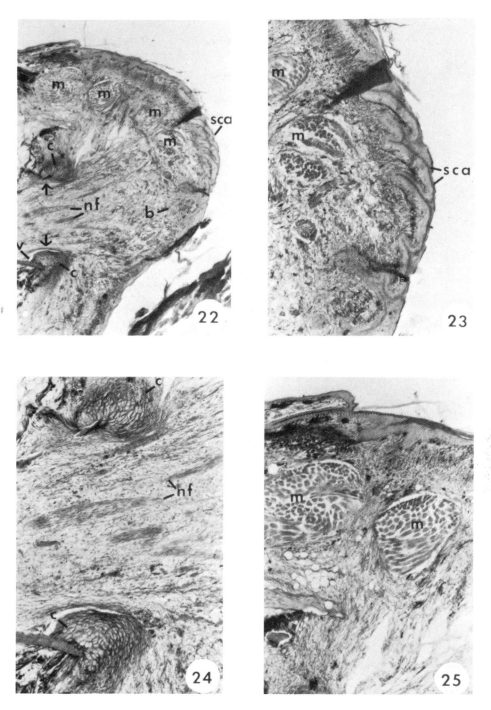

Factors Influencing Rates of Tail Regeneration in the Lizard *Anolis carolinensis*

P. F. A. MADERSON and P. LICHT

In an investigation of the somatotropic effects of certain hormones in the lizard *Anolis carolinensis*, tail regeneration was studied as one of a number of physiological variables related to growth[1]. In spite of a rigidly controlled experimental regime and use of only males of restricted age and size, considerable individual variation in tail regeneration was found. Although such variations have been reported [2,3] they have not been studied in detail. There are conflicting reports regarding certain possible regulatory factors in lacertilian tail regeneration, especially the role of the vertebral autotomy plane, and there have been speculations on largely uninvestigated factors such as epidermal involvement. We attempted to elucidate the basis for individual variation in the regenerative response in *A. carolinensis* by examining these and other factors, especially temperature.

Materials and methods. In early September, 60 adult male *A. carolinensis* (average snout-vent length 64.5 mm, body weight 5.0 g) were put at $32 \pm 0.5\,°C$ with 6 h light daily[4]. Some animals were injected with gonadotropins, or gonadotropins plus prolactin, but there were no significant differences in tail regeneration and the data were pooled for this analysis. Procedures for hand-feeding, assessing growth and autopsy are reported elsewhere[1].

The original tail (average length 124 mm, range 111–138) was amputated with a razor blade 18–21 mm behind the vent: amputated portions averaged 360 mg. None of the animals appeared to have had previously regenerated tails except at the very tip. The length of the regenerating

[1] P. LICHT, Gen. comp. Endocr. *9*, 49 (1967).
[2] S. V. BRYANT and A. D'A. BELLAIRS, J. Linn. Soc. (Zool.) *46*, 279 (1967).
[3] Y. L. WERNER, Acta Zool. *48*, 103 (1967).
[4] P. LICHT, Am. Midl. Nat. *79*, 149 (1968).

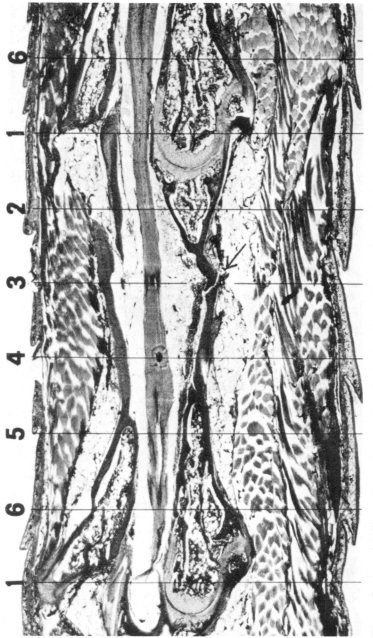

Fig. 1. Section of tail vertebrae of *A. carolinensis* showing the positions used to designate planes of amputation referred to in Figure 3. Although amputations occurred at all positions, when the cut passed through the soft tissues at positions 2 and 4, there was a tendency for the vertebrae to separate at the natural autotomy plane (position No. 3, at arrow).

tissue was measured weekly and after 6 weeks the newly regenerated portion was removed and weighed.

The epidermal condition at amputation and the position of amputation relative to the natural autotomy plane (Figure 1) was determined from histological preparations of the proximal 1.5 cm of the amputated portion: methods are described elsewhere[5].

In order to facilitate comparison between our results and those of previous workers who have used temperatures around 18–22 °C, a second experiment at 21 °C was conducted with 18 males in April. Ad libitum feeding maintained or increased the animals' weights. Severals were transferred to higher temperatures as described below.

Results. Tail regeneration at 32 °C (Figures 2 and 3). No detectable elongation occurred until 7–10 days after amputation, and then there was a period of rapid growth averaging 1.5 mm/day from the 14th to the 28th day. The average growth rate for the 10th to 42nd day after amputation was 0.98 mm/day. The mean length of the regenerated tail at the end of 6 weeks averaged 28.5 mm, representing 28% replacement of amputated tissue. Prominent scales and greatly reduced growth rates during the 6th week suggested that the regenerates were approaching their final length.

The regenerated tails averaged 109 mg (range 45–180 mg) at autopsy, an average replacement of 30%. The final length and weights of the regenerates were highly correlated (r = 0.8, $p < 0.001$). Short tails tended to be proportionally lighter (3 mg/mm) than the long ones (4.3 mg/mm). In general, individual differences in these final values reflected consistent differences in regeneration rates throughout the experimental period.

Despite the relatively uniform rate of blastema formation, rates and final extent of tail regeneration differed markedly among the animals studied.

The following factors were tested against final length and weight of the regenerates: initial and final body weight, change in body weight (all gained weight but this ranged from 1–25%), change in body (snout-vent) length, final weight of the liver and abdominal fat bodies, testicular weight and condition, and thyroid epithelial height at autopsy. None of these variables were significantly correlated ($p > 0.10$) with tail regeneration.

There was no correlation between tail regeneration and the position of amputation with respect to the autotomy plane nor the original epidermal condition (Figure 3).

Tail regeneration at 21 °C. At 21 °C, the time for the first external signs of blastema formation to become evident averaged 36 days, after 35 days in 8 individuals and on the 28th, 39th, 42nd and 45th day in 4 others. This contrasts with an average of about 8 days at 32 °C. There was only

[5] P. F. A. MADERSON and P. LICHT, J. Morph. *123*, 157 (1967).

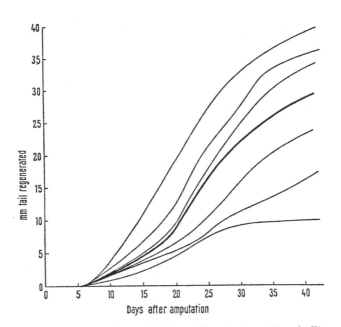

Fig. 2. Tail regeneration at 32 °C in adult male *A. carolinensis*. The central bold curve shows the mean weekly values of 60 animals. The thinner lines depict values for individuals selected to illustrate the variation in regeneration rates.

negligible tail growth during the 2 weeks following the appearance of the blastema. The regenerates grew to 4–7 mm within the next month, averaging 0.15 mm/day. Four animals had regenerated tails of lengths: 3, 11, 14 and 20 mm respectively after 6 months and no increase occurred during a further 4 months.

Three lizards were transferred from 21–32 °C 14 days after amputation. Blastema formation was observed 6 days after the transfer. Thus, there was apparently little progress toward blastema formation in the first 2 weeks at the lower temperature. 5 animals kept initially at 21 °C were transferred to 32 °C after blastema formation. In the following 3 weeks at the higher temperature, tail regeneration proceeded at approximately the same rate – 0.92 mm per day – as in those maintained continuously at 32 °C (see above). In 6 animals transferred from 21–25 °C after blastema formation, the tails grew an average of 6.5 mm (range 2–14 mm) in 3 weeks, i.e. 0.3 mm/day. Tail regeneration effectively ceased when the lizards were returned to 21 °C after 3 weeks at the 2 higher temperatures.

Discussion. The effects of temperature on lizard tail regeneration have not previously been examined quantitatively. Our results indicate that temperature may influence at least 3 different aspects of the regenerative process (Table). Comparison of results obtained between 21 and 32 °C indicate that the higher temperature accelerates both rates of blastema formation and subsequent regeneration rates. However, the latter process is seen to be considerably more temperature dependent than the former when Q_{10} values are compared (Table). Finally, temperature influences the final form of the regenerated tail, a smaller proportion being replaced at the lower temperature. Similar variations in regeneration rates are, however, evident at both temperatures.

The rates of regeneration observed at 21 °C in this study are similar to those previously reported for *A. carolinensis* at this temperature[6-8]. TASSAVA and GOSS[8] reported a replacement of about 23% by length of the amputated tissue after 133 days. The discrepancy between this value and the smaller value observed in our study (17%) may be due to the small sample sizes involved here and to slight differences in the amount of tail amputated (see below). They also used smaller (younger ?) individuals and did not define the sexes. Nevertheless, comparison of their value with that observed at 32 °C further confirms our conclusion that the final length regenerated is temperature dependent.

Possible effects of the individuals' nutritional state and its capacity for tail regeneration were presumably minimized in our study through the use of isocaloric feeding, sufficient to promote some weight gain. Nevertheless, we observed no influence of body growth on regeneration rates despite variations in weight gain between 1% and 25% of initial body weight and linear increases of 0–6 mm. Thus, there appears to be little influence of the nutritional state as long as body weight is maintained; i.e. in the absence of starvation.

[6] R. P. KAMRIN and M. SINGER, J. exp. Zool. *128*, 611 (1955).
[7] J. T. JAMISON, Herpetologica *20*, 145 (1964).
[8] R. A. TASSAVA and R. J. GOSS, Growth *30*, 9 (1966).

Effects of temperature on tail regeneration in *Anolis carolinensis*[a]

Regenerative process	Body temperature		Q_{10}
	21 °C	32 °C	
(1) Time to blastema appearance (days)	36.2	8	3.9
(2) Rate of tail elongation:			
(a) during first month after blastema formation (mm/day)	0.15	0.98	5.5
(b) during 10 days of maximum growth[b] (mm/day)	0.23	1.50	5.4
(3) Final proportion of tail replaced (% tail removed)	17	28	

[a] All values are means for 60 animals. [b] This period begins 20 days after amputation at 32 °C and 76 days after amputation at 21 °C.

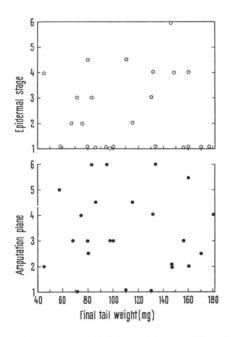

Fig. 3. Relation between final tail weight and epidermal stage at time of amputation (top) and amputation plane within tail (bottom) for 24 of the 60 lizards studied representing the range of regenerative response. The same relationship is seen if final tail length is used. Epidermal stages are based on those described for *A. carolinensis*[5]: stage 1 is a 'resting' stage that lasts 9–10 days following sloughing and stages 2–6 represent the 'proliferative' phase leading to the next molt after another 9–11 days at 32°C. Amputation planes correspond to the positions designated in Figure 1.

Some workers[2,7,9] have reported apparent inhibition or retardation of regeneration following intervertebral amputation. Although inter-familial differences in the importance of this factor may exist, we note that in all 3 studies, the sample sizes were very small. Our data support previous conclusions[3,10,11] that there is no correlation between the position of vertebral breakage and the subsequent rate or extent of regeneration.

The possible role of the epidermis as a modifying influence in tail regeneration is suggested by work attempting to explain the lack of regenerative response in anuran limbs by rapid overgrowth of the epidermis preventing blastema formation[12]. Recent studies of the stratum germinativum of the squamate epidermis throughout the sloughing cycle (PANG, MADERSON and ROTH, unpublished data) have shown peaks of mitotic activity at 3 points in the cycle. However, the observations presented here suggest that the stages with which these peaks are associated[5] do not appear to have any effect on regenerative capacity.

A variety of factors other than those discussed above have been suggested as modifying influences in lacertilian tail regeneration. BRYANT and BELLAIRS[2] and TASSAVA and GOSS[8] demonstrated that rates of regeneration (but not the final proportion regenerated) increased in proportion to the amount of tail removed. HUGHES and NEW[13] and BRYANT and BELLAIRS[2] showed that repeated autotomies lowered regeneration rates. Neither of these factors were involved in the present study, though failure to control them would presumably give rise to variation beyond that reported here.

Field studies have led to suggestions that the age and 'activity' of individuals may modify tail regeneration[14-16]. In view of the wide variations in regeneration obtained under relatively uniform experimental conditions, we suggest that conclusions based on field studies involving small sample sizes may be tenuous.

The peripheral innervation (see references in ZIKA and SINGER[17]) and the ependyma[18] have been shown to be

[9] L. A. MOFFAT and A. D'A. BELLAIRS, J. Embryol. exp. Morph. *12*, 769 (1964).

[10] W. N. F. WOODLAND, Q. Jl microsc. Sci. *65*, 63 (1920).

[11] B. SLOTOPOLSYK, Zool. Jb. Anat. *43*, 219 (1922).

[12] S. M. ROSE, in *Biology of the Amphibia* (Ed. J. A. MOORE; Acad. Press, New York 1964), p. 545.

[13] A. HUGHES and D. NEW, J. Embryol. exp. Morph. *7*, 281 (1959).

[14] H. S. FITCH, Univ. Kansas Publs Mus. nat. Hist. *8*, 1 (1954).

[15] R. E. BARWICK, Proc. R. Soc. N.Z. *86*, 331 (1959).

[16] W. E. BLAIR, *The Rusty Lizard* (Univ. Texas Press 1960).

[17] J. ZIKA and M. SINGER, Anat. Rec. *152*, 137 (1965).

[18] S. B. SIMPSON, in *Regeneration in Animals and Related Problems* (Ed. V. KIORTSIS and H. A. L. TRAMPUSCH; North-Holland Publ. Co., Amsterdam 1965), p. 431.

important determinants of whether regeneration will occur, but their influence on the rates of tail regeneration has not been examined in detail. Such factors may underlie the large degree of variability in our study that was unaccounted for by factors generally thought to influence regeneration rates.

We emphasize the need for care in the selection of sample sizes in tail regeneration studies even when the conditions of the specimens and experimental regime are relatively uniform. The profound temperature effects on tail regeneration raise doubts regarding the validity of interspecific comparisons based on studies where 'comparable' temperatures were not used for all species. In this regard, we suggest that cognizance be taken of the species characteristic preferred body temperatures[4]. These considerations may also be useful in experimental design since regeneration rates most advantageous for study can be attained by selecting appropriate temperatures[19].

[19] Supported by U.S.P.H.S. National Cancer Institute No. 5 RO1 CA 5401-07 and Damon Runyon Foundation DRG 947 (MADERSON) and N. S. F. GB 2885 (LICHT). The authors are indepted to Drs. M. SINGER, S. SIMPSON, S. V. BRYANT and P. COX (Western Reserve University) and Dr. H. WOLFFE (Mass. General Hospital) for comments and advice in the preparation of the manuscript.

THE EFFECTS OF PARTIAL IRRADIATION ON TAIL REGENERATION IN ADULT *SIREDON MEXICANUM*

By V. V. BRUNST, Sc.D.

STUDY of the effects of radiation in the boundary zone of the irradiation field is of especial interest in radiobiology, because those effects may be related to the stimulating or carcinogenic effects of radiation. On the other hand, study of the regeneration process is of interest, not only in biology but also in medicine, because the problem of regeneration may be related to that of cancer. The regeneration process is similar, in some extent, to tumor formation, since both involve intensive local growth in the adult organism. The usual result of regeneration, however, is a replacement of the amputated portion of the body with a normal structure. Even so, the results of regeneration may sometimes include abnormal growth and the formation of tumor-like structures.

Study of the effects of various external factors on the process of regeneration not only adds significantly to our knowledge of the biologic properties of those factors, but also offers valuable information concerning the process of regeneration itself. Regeneration under adverse conditions, for instance, may disclose unusual potentialities of the regeneration blastema, potentialities that cannot be observed under ordinary conditions.

Among the various external factors that can be applied to the regeneration process, radiation is especially useful, because it is able to penetrate deeply into cells, tissues, and organisms without necessarily creating any immediately visible break in natural barriers, and yet it has the capability of inducing the living material that it reaches to undergo appreciable changes in structure or function. Even when it suppresses mitosis, growth, and development, radi-

ation does not always destroy viability; irradiated tissues sometimes live a very long time.[4]

It is a well established fact that irradiation can inhibit regeneration.[1,4,10,16-22] It is also a well established fact that irradiation-induced inhibition of regeneration is permanent. In *Triton cristatus*, for example, the limbs on one side of the body were irradiated locally, then the limbs on both sides were repeatedly amputated during the next 5 or 6 years (those of some animals 3 times, and those of others 4, 5, 6, or even 7 times); regeneration of the unirradiated limbs was completely normal, but the irradiated limbs were never observed to undergo any regeneration whatever.[15]

All of the amputations of irradiated limbs in the study just cited had been irradiated in full. What would have happened in the case of an amputation in a region proximal enough to have been irradiated only in part? Here a broad spectrum of results would be possible, depending upon such factors as the relative number of damaged and undamaged cells, the nature of the damage in the damaged cells, and the severity of that damage.

As a rule, radiation does not affect all cells within the exposed field equally, since the energy supplied by radiation is discontinuous.[25] Thus a very small dose of radiation is sufficient to kill a few of the cells in a specified area of tissue, but a very large dose is necessary to kill all of the cells in that area. The discontinuous effects of radiation are particularly evident along the borders of the irradiated field, where scattered rays strike some cells but not others.

The present investigation is a morphologic study of regenerates that developed from the boundary zone peripheral to the field of irradiation. In the great majority of cases, amputation was within the irradiated field, where regeneration was suppressed completely; but in the small number of remaining cases, amputation was in the boundary zone, where partial irradiation interfered with regeneration to such an extent that all regenerates were more or less monstrous.

MATERIALS AND METHODS

In 2 series of experiments (A-65 and D-65), the tails of 115 adult axolotls (*Siredon mexicanum*) from 2 different spawnings were amputated, and the regenerates that developed were irradiated locally 1 to 2 months later with one or another of the following doses: 1,000 r, 2,000 r, 4,500 r, and 6,000 r. Irradiation conditions were as follows: 250 kv., 30 ma., 0.25 mm. Cu+1 mm. Al filter, beryllium window tube; half-value layer of 0.95 mm. Al; target-to-object distance 20 cm.; and 1,122 r/min. in air.

Before irradiation the animals were anesthetized with a dilute aqueous solution of tricaine methanesulfonate (MS-222), a special agent designed for immobilizing cold-blooded animals, and were kept in water in a plastic tray. Body areas not to be irradiated were shielded by a lead covering made from a sheet of lead 4 mm. thick, with a round radiation localizer 6.5 cm. in diameter.[5] During irradiation the animals were arranged radially within a circle; immediately afterward, they were returned to individual containers supplied with fresh water.

A more proximal amputation of the tails of the 115 animals was performed a relatively long time (495 to 539 days) after irradiation. The material thus obtained was fixed in Stieve's acetic acid solution of mercuric chloride and formaldehyde.[23] After decalcification with 0.5 per cent HCl in 80 per cent alcohol, specimens were embedded in paraffin containing 5 per cent beeswax, and were sectioned at 8 μ. The sections were stained with Ehrlich's hematoxylin and eosin, and photomicrographs were made with most at $\times$ 42, but occasionally at $\times$ 170 or $\times$ 570.

All of these experimental and preparative methods, except those concerned with irradiation, were also applied to 20 unirradiated animals from the same spawnings.

RESULTS

NORMAL REGENERATION OF THE TAILS OF CONTROL ANIMALS

The gradual development of the tail during normal regeneration is shown in Figures 1–3. The tail is already 14 mm. long 26 days after irradiation, but it is most typical for the regenerate to be very wide in the proximal portion; growth of the regenerate occurs from all of the amputation surface (Fig. 1), and hence the proximal width of the regenerate is about 17 mm. The intensity of regeneration is greatest the first month after amputation; the length of the regenerate is 23 mm. after 89 days (Fig. 2), but no more than 25 mm. after 134 days (Fig. 3). Increase in the width of the regenerate, however, is characteristic of this stage of regeneration.

A general picture of the histology of a normal regenerate is provided in Figures 14 and 15. The axial skeleton and the spinal cord are surrounded by a very large region consisting of loose connective tissue, with blood vessels, especially capillaries, but very few muscles. Differentiation of the axial skeleton is highly typical.

A previous study[14] dealt in part with the normal development of the tail of the young axolotl. During the first period (2 or 3 months), development of the axial skeleton of the tail is limited to growth and development of the notochord. The second period begins with the formation of the cartilaginous trunk in the distal portion of the axial skeleton of the tail. The cartilaginous trunk is usually a long and narrow skeletal element without any differentiation during the early stages of development.

In adult tail regenerates fixed 134 days

43

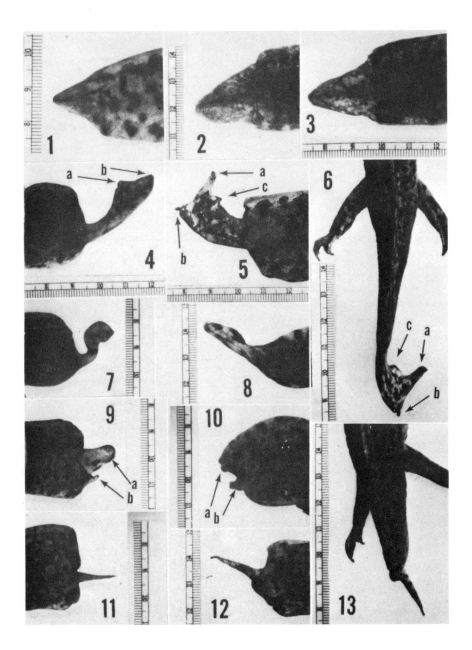

after amputation, rapid differentiation of the cartilaginous trunk, beginning at the actively growing tip, is characteristic. The tip consists of prochondral cells with many mitotic figures (Fig. 17). In the more proximal region (Fig. 16), these prochondral cells are transformed into completely differentiated cartilaginous cells. Four successive zones of the transformation process can be distinguished among the cartilaginous cells (Fig. 16; and 1–4). In the more proximal portions of the cartilaginous trunk, metameric differentiation of cartilaginous tissue, corresponding to the future formation of vertebrae, is highly typical of normal regeneration (Fig. 14; and 15).

In normal regenerates, the spinal cord is characterized by a very monotonous structure in the tail, from the proximal region to the very end. Expansion and duplication of the central canal are never observed in normal regenerates.

THE STRUCTURES OF MONSTROUS REGENERATES
DEVELOPING FROM PARTIALLY
IRRADIATED TAILS

The morphology of the 12 monstrous regenerates exhibited all grades of deviation from normal, ranging from almost normal structures to completely abnormal ones without axial skeleton or spinal cord. Even when similar in outward appearance, some of these regenerates were very different in structure. Furthermore, the cartilaginous trunk did not undergo normal metameric differentiation in any monstrous regenerate in which it developed at all.

The 12 monstrous regenerates can be considered in order of increasing abnormality.

1. A-65-33. A paddle-shaped regenerate (Fig. 4), with the axial skeleton and spinal cord almost normal to the end, without bifurcation. The only abnormality evident, in fact, is absence of normal differentiation of the cartilaginous trunk.

2. D-65-8. A long and narrow regenerate, with the axial skeleton and spinal cord almost normal to the end (Fig. 19).

3. A-65-35. A long, narrow, contorted regenerate (Fig. 7), with the axial skeleton and spinal cord almost normal in the distal portion (Fig. 24), but the cartilaginous trunk abnormally thick in the proximal and middle portions (Fig. 21; and 25).

4. D-65-21. A highly abnormal-looking regenerate, small, awl-shaped, round in cross section (Fig. 12; and 13). The axial skeleton and spinal cord, however, are almost normal to the end (Fig. 18).

←◄◄

PLATE I.

FIG. 1–3. Normal regenerates of control Animals 26 (1), 89 (2), and 134 (3) days after amputation.
FIG. 4. Regenerate of Animal A-65-33 (No. 1) 420 days after second amputation (539 days after irradiation with 1,000 r). a, b, distal ends of regenerate.
FIG. 5 and 6. Regenerate of Animal D-65-23 (No. 11) 365 days after second amputation (495 days after irradiation with 4,500 r). a, b, c, distal ends of regenerate.
FIG. 7. Regenerate of Animal A-65-35 (No. 3) 410 days after second amputation (539 days after irradiation with 1,000 r).
FIG. 8. Regenerate of Animal D-65-12 (No. 6) 365 days after second amputation (495 days after irradiation with 6,000 r).
FIG. 9. Regenerate of Animal A-65-42 (No. 9) 440 days after second amputation (539 days after irradiation with 2,000 r). a, b, distal ends of regenerate.
FIG. 10. Regenerate of Animal D-65-17 (No. 10) 448 days after second amputation (495 days after irradiation with 4,500 r). a, b, distal ends of regenerate.
FIG. 11. Regenerate of Animal A-65-23 (No. 12) 430 days after second amputation (539 days after irradiation with 2,000 r).
FIG. 12 and 13. Regenerate of Animal D-65-21 (No. 4) 381 days after second amputation (495 days after irradiation with 4,500 r).

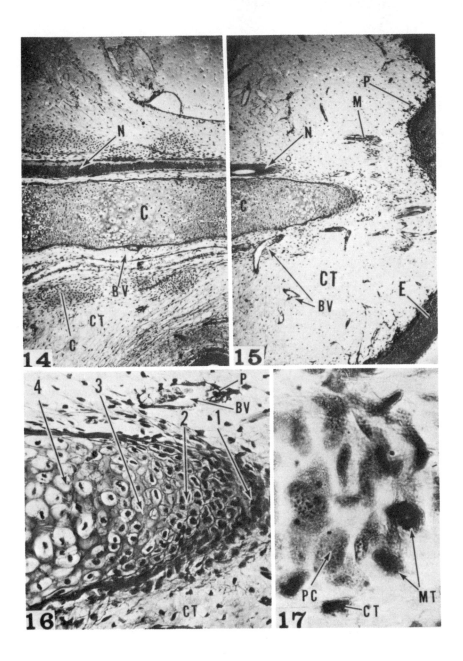

5. D-65-7. A short, wide regenerate, with slight expansion of the cartilaginous trunk and the end of the spinal cord.

6. D-65-12. A long and narrow regenerate (Fig. 8), with considerable expansion of the respective ends of the cartilaginous trunk and the spinal cord.

7. D-65-27. A small and comparatively wide regenerate, with considerable expansion of the respective ends of the cartilaginous trunk and the spinal cord.

8. A-65-36. A small and comparatively wide regenerate, with great expansion of the end of the cartilaginous trunk.

9. A-65-42. A double regenerate, the main regenerate short and comparatively wide and the additional regenerate fairly small (Fig. 9). The regenerate is in the process of reduction. The small regenerate lacks both axial skeleton and spinal cord, and the cartilaginous trunk is completely absent. There is obvious destruction of bony vertebrae by macrophages.

10. D-65-17. A double regenerate, both regenerates very small (Fig. 10). There is expansion of the cartilaginous trunk and the end of the spinal cord. The axial skeleton and the spinal cord do not reach the end of the basal portion of the tail and do not penetrate into either of the regenerates.

11. D-65-23. A comparatively large, paddle-shaped regenerate, with three finger-shaped distal ends (Fig. 5; and 6). The axial skeleton and the spinal cord do not reach the end of the main part of the regenerate and do not penetrate into any of the three ends. There is very great expansion of the end of the cartilaginous trunk, and moderate expansion of the end of the spinal cord.

12. A-65-23. A highly abnormal-looking regenerate, small, awlshaped, round in cross section (Fig. 11), almost identical with No. 4 (D-65-21) in appearance (Fig. 12; and 13), but completely different in structure. The spinal cord and the axial skeleton do not reach the end of the basal portion of the tail and do not penetrate into the regenerate (Fig. 23), which consists only of skin epithelium and loose connective tissue, without any skeleton or spinal cord (Fig. 22).

DISCUSSION

The 12 monstrous regenerates described above represent a wide variety of structures. Even so, one feature is common to all of the monstrous regenerates with axial skeletons, and that is the absence of metameric differentiation of cartilaginous tissue, a process that is highly typical of normal tail regenerates.

Roentgen regression, the destruction of irradiated tissues by macrophages, has been reported a number of times.[1-4,7-11,13-15] Although observed in 90 per cent of regenerates of extremities of adult axolotl during the first 100 to 150 days after irradiation,[10] it was observed in the present study only in one instance.

In addition to the absence of normal differentiation of cartilaginous tissue, the monstrous regenerates described in this paper show several general features:

1. A tendency toward being long but narrow. This tendency is manifested in regenerates ranging from large, paddle-

PLATE II.

Fig. 14–17. Various sagittal sections through normal regenerate of control animal 134 days after amputation. The proximal portion of the regenerate is shown in Figure 14; distal portions are shown in Figures 15–17. BV, blood vessels; C, cartilaginous trunk or other cartilaginous elements; CT, loose connective tissue (or connective tissue cell); E, skin epithelium; M, muscles; Mt, mitoses; N, spinal cord; P, pigment cells; PC, prochondral cells; 1, 2, 3, 4, stages of differentiation of cartilage in distal portion of cartilaginous trunk. (Fig. 14 and 15, ×42; Fig. 16, ×170; Fig. 17, ×570.)

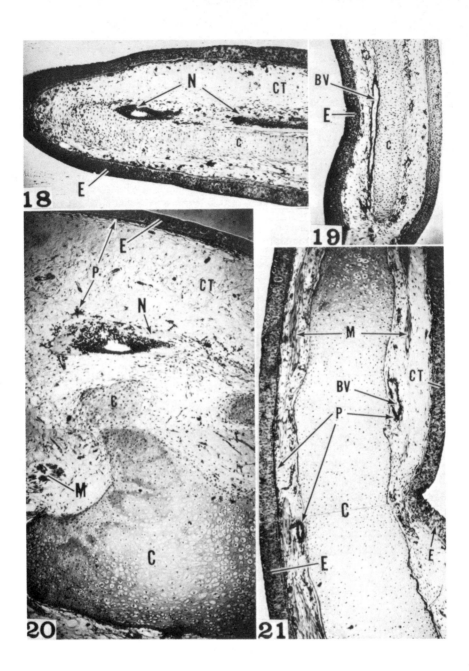

shaped ones (Fig. 4; and 5) to small, narrow, awlshaped ones (Fig. 11; and 13). Only in one instance is there no manifestation of this tendency (Fig. 10).

2. A tendency toward forming several distal ends (Fig. 4; 5; 9; and 10). This tendency is evidently not connected with duplication of the axial skeleton and spinal cord—a phenomenon that was never observed in the present study.

3. A tendency toward preserving a comparatively normal structure, even when the appearance is highly abnormal. The extreme case in No. 4 (D-65-21), a small, awlshaped regenerate, with the axial skeleton and the spinal cord almost normal to the end (Fig. 12; and 13).

4. A tendency toward expansion of the axial skeleton and spinal cord, with inhibition of their growth lengthwise, a tendency manifested only in a few cases. Where this tendency is apparent, the end of the cartilaginous trunk is abnormally thick, and does not reach the end of the basal portion of the tail or penetrate into the regenerate.

5. A tendency toward expansion of only the spinal cord, with inhibition of its growth lengthwise, another tendency manifested only in a few cases. This tendency is associated, in some instances, with transformation of the central canal into a large cavity. Bending of the spinal cord and duplication of the central canal may be accompanied by simultaneous immigration of cells into the cavity (Fig. 23). Similar findings were noted after local irradiation of the tails of young axolotls.[2,14]

Two facts are particularly noteworthy: first, despite their great variety in outward appearance, several of the monstrous regenerates are almost normal in structure; and second, despite their tendency toward forming multiple ends, none of the monstrous regenerates exhibit duplication of the axial skeleton and spinal cord. The tendency toward lengthwise growth is the most important stimulus in the regeneration of the amphibian tail. If the regeneration blastema includes enough cells that are undamaged or only slightly damaged, impairment of regeneration is only partial, and the most important portions of the regenerate—the axial skeleton and the cord—are almost normal to the end of the regenerate.

It is not a coincidence that the monstrous regenerates are long and narrow in almost all cases. In a radiation-damaged regeneration blastema, presumably the axial skeleton and the spinal cord are formed first. If no more than a limited number of undamaged or only slightly damaged cells are available, the upper and lower portions of the tail are not formed at all. Only in such a way can the formation of awlshaped regenerates (Fig. 11; and 12) be explained.

If enough undamaged material is available, the axial skeleton and the spinal cord

←‒‒

PLATE III.

Fig. 18. Distal portion of regenerate of Animal D-65-21 (No. 4) 381 days after second amputation (495 days after irradiation with 4,500 r). (Same as in Fig. 12 and 13.) C, cartilaginous trunk; CT, loose connective tissue; E, skin epithelium; N, spinal cord. (×42.)

Fig. 19. Distal portion of regenerate of Animal D-65-8 (No. 2) 370 days after second amputation (495 days after irradiation with 6,000 r). BV, blood vessel; C, cartilaginous trunk; E, skin epithelium. (×42.)

Fig. 20. Distal portion of regenerate of Animal D-65-23 (No. 11) 365 days after second amputation (495 days after irradiation with 4,500 r). C, abnormal cartilaginous trunk; CT, loose connective tissue; E, skin epithelium; M, muscles; N, distal portion of spinal cord; P, pigment cells. (×42.)

Fig. 21. Middle portion of regenerate of Animal A-65-35 (No. 3) 410 days after second amputation (539 days after irradiation with 1,000 r). (Same as in Fig. 7.) BV, blood vessel; C, abnormal cartilaginous trunk; CT, loose connective tissue; E, skin epithelium; M, muscles; P, pigment cells. (×42.)

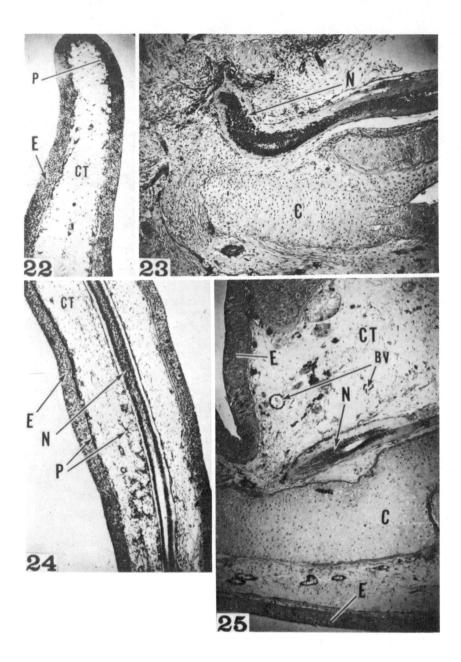

can obviously be normal and reach the end of the regenerate, as is the case with No. 4 (D-65-21), shown in Figures 12 and 13. If the supply of undamaged material is sufficiently limited, however, the regenerate may consist only of skin epithelium and loose connective tissue and lack any axial skeleton, as is the case with No. 12 (A-65-23), shown in Figure 11. It seems evident that this last case is highly exceptional; even in the absence of enough undamaged material, the regeneration process is very unlikely to involve destruction of the most important parts of the regenerate.

mations with almost normal axial skeletons and spinal cords to small and narrow awl-shaped formations lacking both axial skeletons and spinal cords. Regression of the axial skeleton, which has been observed in previous study in 90 per cent of regenerates of axolotl extremities during the first 100 to 150 days after irradiation, was observed in only one instance in the present study. Pathologic changes in the distal portion of the spinal cord were minor and very similar to such changes in irradiated tails of young axolotls, but ingrowth of the spinal cord in the distal portion of the regenerate did not occur in some cases.

SUMMARY

1. The tails of 115 adult axolotls (*Siredon mexicanum*) were amputated, and the regenerates that developed were irradiated locally 1 to 2 months later with one or another of the following doses: 1,000 r, 2,000 r, 4,500 r, and 6,000 r. A more proximal amputation, performed a relatively long time (more than 1 year) after irradiation, was followed by the development of only 12 regenerates, all monstrous. Obviously these regenerates developed from the boundary zone, where irradiation of the cellular material had been only partial. In the other 103 animals, regeneration was completely suppressed.

2. The 12 monstrous regenerates were the products of limited regeneration, and represented various types of abnormalities, ranging from narrow paddle-shaped for-

REFERENCES

1. BRUNST, V. V. Untersuchung des Einflusses von Röntgenstrahlen auf die regenerierenden und erwachsenen Extremitäten bei Urodelen. *Arch. Entwicklungs Mechn. d. Organ*, 1944, *142*, 668–705.
2. BRUNST, V. V. Effects of local x-ray irradiation on tail development of young axolotl. *J. Morphol.*, 1950, *86*, 115–151.
3. BRUNST, V. V. Influence of local x-ray treatment on development of extremities of young axolotl (Siredon mexicanum). *J. Exper. Zool.*, 1950, *114*, 1–49.
4. BRUNST, V. V. Influence of x-rays on limb regeneration in urodele amphibians. *Quart. Rev. Biol.*, 1950, *25*, 1–29.
5. BRUNST, V. V. Technics of local low voltage roentgen ray irradiation of experimental animals. *Lab. Invest.*, 1952, *1*, 432–438.
6. BRUNST, V. V. Effect of different doses of roent-

←⬚

PLATE IV.

FIG. 22. Distal portion of regenerate of Animal A-65-23 (No. 12) 430 days after second amputation (539 days after irradiation with 2,000 r). (Same as in Fig. 11.) This regenerate has no axial skeleton or spinal cords CT, loose connective tissue; E, skin epithelium; P, pigment cells. (×42.)

FIG. 23. Proximal portion of same regenerate as in Figure 22. Distal portions of axial skeleton (C) and spinal cord (N). (×42.)

FIG. 24. Distal portion of regenerate of Animal A-65-35 (No. 3) 410 days after second amputation (539 days after irradiation with 1,000 r). (Same as in Fig. 7.) CT, loose connective tissue; E, skin epithelium; N, spinal cord; P, pigment cells. (×42.)

FIG. 25. Proximal portion of same regenerate as in Figure 24. BV, blood vessel; C, cartilaginous trunk; CT, loose connective tissue; E, skin epithelium; N, spinal cord. (×42.)

gen rays on adult axolotl (Siredon mexicanum). AM. J. ROENTGENOL., RAD. THERAPY & NUCLEAR MED., 1958, *80*, 126–142.

7. BRUNST, V. V. Roentgen sensitivity of various portions of eye of young axolotl (Siredon mexicanum). AM. J. ROENTGENOL,, RAD. THERAPY & NUCLEAR MED., 1958, *80*, 1014–1030.

8. BRUNST, V. V. Roentgen regression and roentgen stimulation in axolotl (Siredon mexicanum). *Acta Unio internat. contra cancrum*, 1959, *15*, 568–576.

9. BRUNST, V. V. Effect of local x-ray irradiation upon teeth of adult axolotl (Siredon mexicanum). *J. Dent. Res.*, 1959, *38*, 301–310.

10. BRUNST, V. V. Reaction of limb regenerates of adult axolotl (Siredon mexicanum) to x-irradiation. *Radiation Res.*, 1960, *12*, 642–656.

11. BRUNST, V. V. Roentgen regression in axolotl (Siredon mexicanum). AM. J. ROENTGENOL., RAD. THERAPY & NUCLEAR MED., 1961, *85*, 158–178.

12. BRUNST, V. V. New observations concerning roentgen sensitivity of pigment cells in young axolotls (Siredon mexicanum). AM. J. ROENTGENOL., RAD. THERAPY & NUCLEAR MED., 1965, *93*, 222–237.

13. BRUNST, V. V. Effects of ionizing radiation on development of amphibians. *Quart. Rev. Biol.*, 1965, *40*, 1–67.

14. BRUNST, V. V. Histopathology of development of axial skeleton and spinal cord in irradiated tail of young axolotl (Siredon mexicanum). AM. J. ROENTGENOL., RAD. THERAPY & NUCLEAR MED., 1965, *95*, 992–1012.

15. BRUNST, V. V., and SHEREMETEVA, E. A. Sur la perte locale du pouvoir régénérateur chez le triton et l'axolotl causée par l'irradiation avec les rayons X. *Arch. Zool. expér. et génér.*, 1936, *78*, 57–67.

16. BUTLER, E. G. Effects of x-radiation on regeneration of fore limb of amblystoma larvae. *J. Exper. Zool.*, 1933, *65*, 271–315.

17. BUTLER, E. G., and O'BRIEN, J. P. Effects of localized x-radiation on regeneration of urodele limb. *Anat. Rec.*, 1942, *84*, 407–413, 466–467.

18. BUTLER, E. G., and PUCKETT, W. O. Studies on cellular interaction during limb regeneration in amblystoma. *J. Exper. Zool.*, 1940, *84*, 223–237.

19. LITSCHKO, E. J. Observations sur la régénération des extremités des axolotls après l'action des rayons X. *C. R. Acad. Sci. U.S.S.R.*, 1930, *A*, 594–596.

20. LITSCHKO, E. J. Einwirkung der Röntgenstrahlen auf die Regeneration der Extramitäten, des Schwanzes und der Dorsalflosse beim Axolotl. *Trav. Lab. Zool. Exper. Acad. Sci. U.S.S.R.*, 1934, *3*, 101–140.

21. PUCKETT, W. O. Effects of x-radiation on limb development in amblystoma. *Anat. Rec.*, 1934, Suppl. 58, 32–33.

22. PUCKETT, W. O. Effects of x-radiation on limb development and regeneration in amblystoma. *J. Morphol.*, 1936, *59*, 173–213.

23. ROMEIS, B. Mikroskopische Technik. Leibniz Verlag, München, 1948, p. 74.

24. UMANSKI, E. Untersuchung des Regenerationsverganges bei'Amphibien. *Biol. Zeitschr.*, 1937, *6*, 739–756.

25. WARREN, S. Histopathology of radiation lesions. *Physiol. Rev.*, 1944, *24*, 225–238.

Effects of Amputation of Limbs and Digits of Lacertid Lizards

A. d'A. BELLAIRS AND SUSAN V. BRYANT

Amputation of digits has often been practiced by herpetologists as a method of recognizing individuals in field and laboratory studies. While there is a general impression that such digits do not grow again, Marcucci ('30) described the appearance of a small outgrowth with a claw at its tip in one specimen of *Lacerta muralis*. It therefore seemed of interest to follow the histological changes which occur after amputation of the digits of lizards and to compare them with those seen after limb amputation, a procedure which is occasionally, though not normally, followed by regeneration (Marcucci, '30; Barber, '44). Attempts to determine the effects of limb amputation during the later stages of embryonic life have also been made.

MATERIALS AND METHODS

Amputations of the limbs of adult lizards were made through the middle of one thigh with a razor blade, under ether anaesthesia, in nine individuals of *Lacerta vivipara* and two of *Lacerta dugesii*, the Madeira wall lizard. Adult digits, usually the second, third or fourth, were clipped off with scissors through the second or third phalanx in five *L. vivipara* and ten *L. dugesii*. Serial sections of the stumps, fixed at various intervals after injury, were stained with Heidenhain's azan. The *Lacerta vivipara* were kept in the laboratory at a temperature of about 27°C. The *L. dugesii* were kept slightly warmer throughout, at 27–30°C.

Limb amputations were performed on 120 embryos of *L. vivipara* after the eggs had been removed from the mother and placed in Panigel culture (Moffat and Bellairs, '64). One hind limb was removed close to the body with a fine pair of scissors, after the shell membrane, chorio-allantois and amnio-allantois had been incised. In some cases incisions were made in the membranes, without the limb of the embryo being amputated. After operation, the embryos were allowed to develop for various periods at 28°C before being fixed in Bouin. Serial sections of the pelvic regions of the embryos were prepared and stained with Heidenhain's azan.

RESULTS

1. *Amputation of limbs in adults.* After the operation the limb usually bleeds for a few minutes and the lizards generally drink freely if they are placed partly in a

[1] Supported as a research assistant by grants from the Medical Research Council to Dr. A. d'A. Bellairs.

dish of water when they have recovered from the anaesthetic. Removal of one hind limb has little effect on locomotion. The following observations refer to *Lacerta vivipara* except when stated.

In a lizard examined one week after operation the area of the wound was much the same size as immediately after injury, but was covered by a scab. Microscopic examination shows that new epidermis has grown out from the hinge regions of the scales bordering the wound and already extends some distance beyond its margins (fig. 1). This epidermis is greatly thickened, consisting in places of some 12 layers of cells, more than twice the number seen in normal epidermis. These are covered by several irregular layers of keratin, some of which have become detached in preparation. In places the severed muscles are undergoing extensive dedifferentiation and liberated cells are accumulating as a blastema beneath the new epidermis. The cut end of the femur projects slightly from the surface of the wound. A second lizard examined ten days after operation shows an almost healed limb stump. Here the shaft of the femur has been eroded by osteoclasts, and the epidermis has grown into the excavated bone. Presumably the distal portion of the bone would have been sloughed in time, and the advancing edges of the epidermis would have met in the center of the wound.

In a lizard examined after two weeks the area of the lesion is almost covered by new epidermis, but a large space filled with blood cells and covered by the remains of a scab is present in its central part which is still bare. The cut femur ends some distance beneath this, suggesting that a portion of the bone has either been cut off by epidermal penetration and sloughed, or has been absorbed. A dense accumulation of dedifferentiated blastema cells has formed around and distal to the end of the femur, and multinucleated giant cells are present at the surface of the bone. A specimen examined after three weeks is very similar except that a nodule of cartilage is forming near the cut end of the femur.

In one four-week specimen, as in that of ten days previously described, the epidermis is growing into the eroded shaft of the femur (fig. 2), while in another of the same age the wound is entirely covered by thickened epidermis which has not yet developed new scales; it is impossible to tell whether any part of the bone has sloughed. In both specimens large masses of cartilage have formed along the sides of the femur on the proximal aspect of the new epidermis. Blastema cells are less evident than in the previous specimens.

In two further lizards examined after six and eight weeks respectively the stumps of the limbs have become smooth and rounded. The new epidermis is thicker than normal and is not elevated to form scales. No dermal pigment cells are present beneath it; the band of dermal pigment ends abruptly at the margins of the original wound site in such a way as to suggest that no substantial contraction of the tissues has taken place. A massive investment of cartilage encases the cut end of the femur and continues for some distance up the shaft like a sleeve. Connective tissue intervenes between this cartilage and the epidermis.

In one specimen of *Lacerta vivipara* examined after 12 weeks (fig. 3) and in two *L. dugesii* after 13 weeks further differentiation of the limb tissues has taken place. New scales are beginning to form in the terminal epidermis and the cartilage which surrounds the femur has become calcified.

Portions of cartilage tube containing ependyma were taken from the mature tail regenerates of the two *Lacerta dugesii* and implanted into their limb stumps immediately after amputation. These implants did not have the effect of inducing regenerative changes in the limbs similar to those described by Simpson ('64) in the tail of the skink *Lygosoma*. In one *L. dugesii* the piece of cartilage tube remained readily identifiable and still retained a portion of ependymal tube, the structure regenerated in lieu of normal spinal cord (see Hughes and New, '59). In other respects the appearances were very similar to those in the 12-week limb stump of *Lacerta vivipara*. It does not seem possible that any of these specimens would have regenerated their limbs if they had been allowed to survive for longer, since the tissues at the tip of the stump are well differentiated.

2. *Amputation of digits in adults.* These observations were made on specimens of *L. vivipara* and *L. dugesii.*

Three specimens which were examined one to two weeks after injury, show the wound area covered by a thick scab which contains a separate fragment of the amputated phalanx; presumably this had been cut off by epithelial penetration combined with osteoclastic activity, like the ends of some of the femora after limb amputation.

In two specimens examined three weeks after amputation, the scab has disappeared and the wound is covered completely with new epidermis. In the specimen illustrated in figure 4, this epidermis is thicker than that over the rest of the digit, and has several layers of irregular keratin. Osteoclasts are numerous around the cut end of the phalanx which appears to be in the process of absorption. Many cells derived from the various tissues at the site of the wound have accumulated beneath the new epidermis; around the shaft of the bone a sleeve of cartilage has been formed. The appearance of this specimen is in particular very reminiscent of that of an early tail regenerate with blastema and an apical cap of thickened epidermis. The papilla growing proximally from the epidermis into the blastema is remarkably like the structure which has been described by Hughes and New ('59) in early tail regenerates of geckos.

In four specimens examined four to six weeks after injury the epidermis over the wound is still much thicker than normal, but no further growth of regenerative type beyond the rounding off of the cut end of the digit has taken place (fig. 5). Blastema cells are less evident than at three weeks, and the phalanx is now encased by cartilage and connective tissue which extends proximally almost as far as the next interphalangeal joint.

Five specimens examined after 9 and 15 weeks show that the new skin over the wound has become very similar to that of the adjacent sides of the digit. The blastema has now disappeared, and the bone is completely invested with cartilage. Many osteoclasts are present, and it is possible that resorption of the whole remaining part of the phalangeal shaft would

eventually have taken place. None of these specimens had regenerated even a small part of the tissue which was amputated.

3. *Amputation of limbs in embryos.* Attempts were made to determine the regenerative capacity of the hind limb of *Lacerta vivipara* by amputating the appendage at various later stages in embryonic life. As reported by Marcucci ('15) in the oviparous *Lacerta muralis*, no regeneration was observed. The epidermis closes over the wound surface within two to three days, and if the specimen is allowed to survive for long enough, scales much like those elsewhere form on the limb tip. As in adult limbs and digits after amputation, cartilage is formed around the femur of embryos operated on during the later stages of embryonic life. Unfortunately, operations on lizard embryos are often followed by the development of constriction bands which may cause auto-amputation of an appendage such as the tail or limb. These may be extremely difficult to detect and the final result may appear similar to that of the experimental injury. Control experiments in which the amnion was incised but the embryo left intact were followed in a few cases by such auto-amputation. The effect of this self injury may be to inhibit regeneration, since amputation is brought about by constriction.

Although it seems likely that the majority of our specimens show the true result of experimental injury, we cannot with certainty state that the limb of the embryo is unable to regenerate.

DISCUSSION

The process of healing after amputation is essentially similar in both the hind limbs and digits of adult lizards. A notable feature is the great thickening of the new epidermis; this was also observed in the embryos and seems to be a characteristic reptilian reaction to injury of the type inflicted. The extensive formation of cartilage which spreads far beyond the immediate neighborhood of the wound is also interesting. It was observed by Barber ('44) after amputation of the forelimbs of *Anolis,* and by Pritchard and Ruzicka ('50) in the callus formed after experimental fractures of the long bones of *Lacerta vivipara.* The latter authors commented on the predominance of cartilage in healing bones of cold-blooded vertebrates (frog and lizard) as compared with conditions in the rat.

The ultimate fate of this cartilage in our lizards was not determined, and it was still present as long as three months after injury. It is possible that it would eventually have become ossified and then remodeled so that the swollen shaft of the bone would have reverted to its original shape.

The early stages of healing of the limbs and digits show many features in common with those of early tail regenerates. In both cases a substantial blastema is accumulated beneath a thickened wound epidermis. Dedifferentiation of muscle in the limb, however, is very extensive whereas in the autotomized tail the integrity of the muscle is preserved by the fibrous autotomy septum so that this tissue contributes little if at all to the blastema. In both limbs and digits there is a tendency for a small portion of the cut end of the femur or phalanx to become cut off by osteoclastic activity and sloughed soon after injury. Similar sloughing of a part of the remaining fragment of the autotomized caudal vertebra has been described in geckos by Werner ('67), but we have found no definite evidence of this in *Lacerta vivipara*.

Despite the fact that the limbs of *Lacerta* occasionally regenerate in part, the stumps of amputated limbs show a less regenerative appearance than those of the digits which we have examined. This appearance is particularly striking in the digit shown in figure 4 which has many of the characters seen in some tail regenerates, including an epidermal papilla. It is of course possible that this digit would have regenerated if it had been left for a longer period, but conditions in other specimens suggest that the early promise of regenerative capacity is not fulfilled. As in the healing limbs the epidermis does not remain thickened and eventually reverts to its normal state. The blastema does not increase in size and most of its cells become converted into cartilage around the bone.

Barber ('44) compared the results of amputation through the forearm in *Anolis* with those of breaking off the tail. She attributed the failure of the limb to regen-

erate to the fact that its tissues are more loosely packed than those of the tail and so permit much greater shrinkage after injury. We have not observed such shrinkage in our experiments. Moreover, we have found little evidence for the contraction of wounds which is characteristic of healing of not only excised skin wounds (Billingham and Medawar, '55; James, '64), but also of amputated digits (Schotté and Smith, '59) in mammals. In the limb stumps of *Lacerta* which were sectioned 12 or 13 weeks after injury the greater part of the wound is covered by new epidermis rather than by the approximated edges of the original skin. This also appears to be the case in a few reptiles which we have examined which had old injuries sustained before capture.

Our observations suggest that shrinkage is not a critical factor in determining the regenerative capacity of an amputated appendage. The structure of the digits which have very little soft tissue between their rigid scaly covering and the bone of the phalanx would seem to preclude shrinking, and little if any was observed to occur. The balance between healing and regeneration is probably determined by other factors, for example, as Zika and Singer ('65) have demonstrated in *Anolis*, the quantity of nervous tissue present in the region of the wound.

The limbs of a considerable number of lizards were amputated during the later stages of embryonic life, and in no case case was regeneration observed. However the complicating effects of the embryonic membranes make it impossible to be certain that the limb of the embryo is devoid of regenerative capacity.

LITERATURE CITED

Barber, L. W. 1944 Correlations between wound healing and regeneration in forelimbs and tails of lizards. Anat. Rec., 89: 441–453.

Billingham, R. E., and P. B. Medawar 1955 Contracture and intussusceptive growth in the healing of extensive wounds in mammalian skin. J. Anat. 89: 114–123.

Hughes, A., and D. New 1959 Tail regeneration in the geckonid lizard, *Sphaerodactylus*. J. Embryol. exp. Morph., 7: 281–302.

James, D. W. 1964 Wound contraction—a synthesis. In: Advances in Biology of Skin, 5. Pergamon Press, London.

Marcucci, E. 1915 Gli arti e la coda della *Lacerta muralis* rigenerano nello stadio embrionale? Boll. Soc. Nat. Napoli, 27: 98–101.

————— 1930 Il potere rigenerativo degli arti nei rettili. Archivio Zoologico, *14:* 227–252.

Moffat, L. A., and A. d'A. Bellairs 1964 The regenerative capacity of the tail in embryonic and post-natal lizards (*Lacerta vivipara* Jacquin). J. Embryol. exp. Morph., *12:* 769–786.

Pritchard, J. J., and A. J. Ruzicka 1950 Comparison of fracture repair in the frog, lizard and rat. J. Anat., *84:* 236–261.

Schotté, O. E., and C. B. Smith 1959 Wound healing processes in amputated mouse digits. Biol. Bull., *117:* 546–561.

Simpson, S. B. 1964 Analysis of tail regeneration in the lizard *Lygosoma laterale.* I. Initiation of regeneration and cartilage differentiation: The role of ependyma. J. Morph., *114:* 425–436.

Werner, Y. L. 1967 Regeneration of the caudal axial skeleton in a gekkonid lizard (*Hemidactylus*) with particular reference to the 'latent' period. Acta zool., Stockh., *48:* 103–125.

Zika, J., and M. Singer 1965 The relation between nerve fiber number and limb regenerative capacity in the lizard, *Anolis.* Anat. Rec., *152:* 137–140.

Abbreviations

B, blastema; C, cartilage; DP, band of dermal pigment cells; E, epidermis; EP, epidermal papilla; F, femur; MU, muscle; PH, phalanx; UN, small undifferentiated cells.

PLATE 1

EXPLANATION OF FIGURES

1 Limb stump of adult *Lacerta vivipara* one week after amputation. At each side of the wound, the epidermis between the arrows has migrated over the underlying tissues. The femur and some muscles project from the surface of the wound. Undifferentiated cells are accumulating beneath the epidermis.

2 Almost healed limb stump of adult *L. vivipara* four weeks after amputation, the femur projects beyond the level of the epidermis, its shaft is being eroded and penetrated by migrating epidermis and osteoclasts. Some undifferentiated cells are present beneath the new epidermis, but around the shaft of the femur, they have differentiated into cartilage.

3 Limb stump of *L. vivipara* 12 weeks after amputation. The new epidermis is still distinguishable from that elsewhere on the limb since scales are not fully differentiated and it does not possess an underlying layer of dermal pigment cells. A massive cartilage cap can be seen over the end of the femur.

4 Rounded stump of a digit of *L. dugesii* three weeks after amputation. Beneath the thickened epidermis which covers the tip of the digit a blastema of undifferentiated cells has accumulated. This is penetrated by a downgrowth of the epidermis, the epidermal papilla. Around the shaft of the phalanx, which appears to be undergoing resorption, a cartilage sleeve can be seen.

5 Stump of a digit of *L. vivipara* four weeks after amputation. A thickened layer of epidermis covers the tip of the stump. The cut end of the phalanx is sealed by cartilage, which also extends for some distance proximally along the shaft.

58

PLATE 1

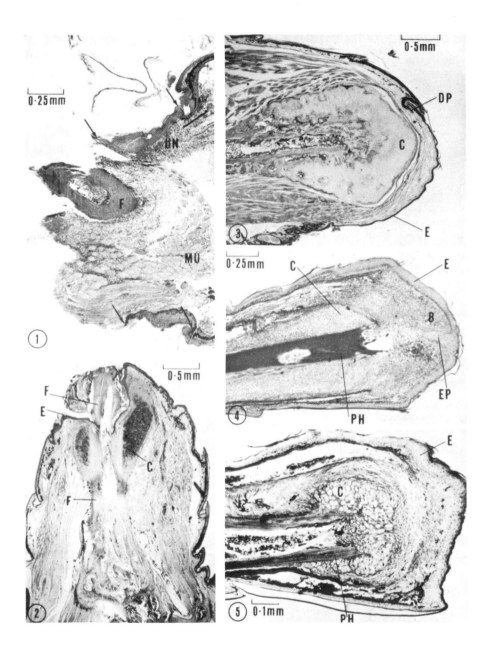

Further Observations on Tail Regeneration in
Anolis carolinensis (Iguanidae, Lacertilia)

P. F. A. MADERSON AND S. N. SALTHE

Maderson and Licht ('68) reported that the rate and final extent of tail regeneration varied greatly among 60 adult male *Anolis carolinensis* maintained under closely controlled experimental conditions. Attempts to explain this variation by correlation with a variety of physiological parameters, with original size and growth changes, with position of amputation within a vertebra, and with epidermal activity at the time of operation all proved unsuccessful. A number of possible explanations for the reported variation could not be examined in the original experiment. First, the phenomenon might have a genetic basis representing either differential regenerative capacities in different natural populations, or between individual animals. Second, since only adults were examined previously, the variation might be associated with age. Third, it was possible that the specific constant conditions of maintenance used in the original experiment were responsible for the variation, in that they were in some way different from conditions in which the species is able to perform normally.

The present work reports the results of experiments designed to test the above-mentioned possibilities. In addition, the implications of variation in tail regeneration in this species will be discussed.

MATERIALS AND METHODS

In April 1969 male *A. carolinensis* collected within a five mile radius were obtained from a commercial supplier in Louisiana. Forty adults (average snout-vent length 62 mm, body weight 4.57 gm) and 40 juveniles (average snout-vent length 48 mm, body weight 2.38 gm) without previously regenerated tails were selected. The tails were amputated with a razor blade 21 mm behind the vent in the adults (average length of tail removed 84 mm), or 16 mm behind the vent in the juveniles (average length of tail removed 64 mm). The animals were divided into groups and maintained under different regimes as follows.

Group A ("Natural"). Twenty adults and 20 juveniles were placed in separate cages in a south-easterly facing window. The interior aspect of the cage always had a minimum temperature of 15°C. Solar heat and light varied considerably throughout the course of the experiment (April-August), but on "dull" days, heat and light were augmented by electric lamps. For reasons that will be discussed later, the environmental conditions were deliberately allowed to vary considerably.

Group B. Ten adults and ten juveniles were placed in separate cages inside an incubator at 31.5 ± 1°C with a photoperiodicity of eight hours light, 16 hours dark. A fluorescent tube was used.

Group C. Ten adults and ten juveniles were placed in similar incubator conditions to Group B but with a photoperiodicity of eight hours dark, 16 hours light.

All three groups were fed *ad libitum* on crickets, mealworms, and *Drosophila* flies. Water was provided *ad libitum* in dishes and by daily spraying.

Tail regeneration was measured weekly from the twelfth day after amputation on.

Seven weeks after amputation, the animals were weighed and the snout-vent length measured. The tails were amputated a second time 18 mm and 13 mm behind the vent in the adults and juveniles respectively. Again, tail regeneration was measured weekly for a nine week period for all groups, at which time the incubator studies were ended. Many individuals from Group A were observed for periods up to eight months following the second amputation.

Most of the animals survived the initial 16 weeks of the experiment; their feeding appeared normal; all maintained or gained weight; and continued their normal shedding periodicities.

Since the purpose of this study was solely to determine the existence or non-existence of variability, we had no need to compare one group with another statistically. We have chosen to use the coefficient of variability (V) despite Lewontin's ('66) demonstration that a logarithm of the variance is a more powerful means of handling variability. Since our points could be made retaining the conventional and more widely-known method of expressing variability, we felt that communication could be enhanced by so-doing.

RESULTS

General comments. The general pattern of regenerative response in all groups, for both first and second amputations was essentially similar to that reported by Maderson and Licht ('68). Measurable growth in many juveniles could not be detected until 12-14 days after amputation, and there was evidence of some delay in blastema formation in some adults in Group A. Maximum growth rate periods following the initial refractory period were observed in all groups, but were less obvious during the second regeneration (fig. 1). Regeneration then slowed considerably during the seventh week after the first amputation, but the flattening of the growth curve was slightly less acute for the second period. Some individuals in Group A maintained for over 200 days following the second amputation showed 1-2 mm further growth of the regenerate after the initial seven week curve indicated in figure 1.

Variation. Data for 67 animals surviving seven weeks after the first amputation, and a further seven weeks after the second amputation are shown in figure 2. Variation in individual response was found in both age groups under all environmental conditions. The phenomenon of large individual variation in each regenerative period is illustrated by the sample standard deviations within the groups shown in figure 1. Coefficients of variability for each group are shown in table 1. The least variation (V = 25) was seen in the first regeneration in adults maintained at 32°C with eight hours light daily; the greatest (V = 73) was seen in the second regeneration for juveniles kept under "natural" conditions.

Individual regenerative responses. In general, individuals that started their measurable growth responses earlier or later than the average maintained their standard of performance throughout one regenerative period, although some exceptions were noted. However, there was no

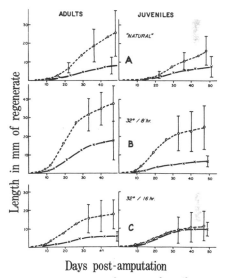

Days post-amputation

Fig. 1   Average growth curves for tail regeneration in male adult and juvenile *A. carolinensis* maintained under various experimental conditions. The codes describing the various groups A-C are given on page 185. First regeneration is indicated,- - - -; second regeneration is indicated as, —x—x—. Vertical solid lines represent one standard deviation above and below the mean at various times throughout the experiment.

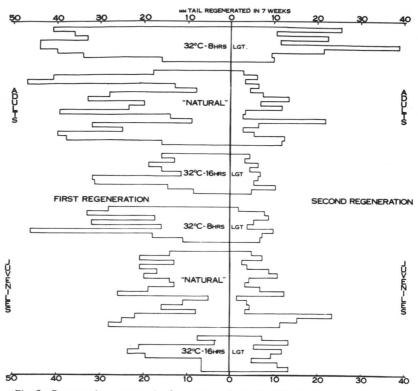

Fig. 2 Regenerative response in the seven week periods following two successive amputations of the tail in adult and juvenile male *A. carolinensis* maintained under various environmental regimes (codes given on page 185). The performances of any one animal in the two periods may be compared by placing a horizontal straight edge between two opposite short vertical lines to the left and right of the median axis 0-0.

evidence of consistent repetition of "good" or "poor" or "average" response in the two consecutive regenerations.

Effects of environment. In general, the three experimental regimes used affect regenerative response of both adults and juveniles in the same way for the first regeneration. Maximum regeneration was seen at 32°C with eight hours light, minimum at 32°C with 16 hours light, with an intermediate situation observed in the "natural" animals (Group A).

Effects of consecutive amputations. The values and shape of the growth curves (figs. 1, 2, and table 1) indicate a depression of mean regenerative response in all groups following the second ampu-

tation, in spite of a few individual exceptions (fig. 2). For the juveniles maintained at 32°C with 16 hours light per day, the figures are too small for accurate analysis. For the other groups, 16-44% of the *original* tail removed is replaced during the first regeneration (table 2, column C), but only 7-21% of the *total original* tail removed is replaced seven weeks after the second amputation (table 2, column G). However, if regenerative capacity is defined as the percentage of replaced material following first *or* second amputation (e.g., the average length of "tail" removed at the second amputation consisted of a few millimeters of normal tail, plus the average result of the first regenerative

TABLE 1

Variability in anolis tail regeneration

	First regeneration				Second regeneration			
	n	Mean	V	Days	n	Mean	V	Days
		mm				mm		
Natural adults	19	25.8	47	48	17	7.8	65	51
Natural juveniles	19	16.3	42	48	17	7.7	73	51
32°C/8 hour adults [1]	9	37.2	25	47	8	18.8	55	49
32°C/8 hour juveniles	9	25.2	46	47	8	6.6	39	49
32°C/16 hour adults	9	18.0	43	47	9	5.9	39	49
32°C/16 hour juveniles	9	11.8	64	47	8	9.8	36	49
M & L, '68 32°C/6 hour [1]	60	28.7	24	42				

The codes describing the various experimental groups are given on page 185.
[1] The V value for the nine animals in the 32°C/8 hour group in this study may be compared with that for the 60 animals reported by Maderson and Licht ('68).

TABLE 2

	A Average length of tail removed first amputation	B Average length of first regenerate	C Replacement in first regenerate $\frac{B}{A} \times 100$	D Average length tail removed second amputation	E Average length of second regenerate	F Replacement in second regenerate $\frac{E}{D} \times 100$	G Original tail replaced in regeneration $\frac{E}{87 \text{ (or 67)}} \times 100$
	mm	mm	%	mm	mm	%	%
"Natural" adults	84	25.8	31	28.8	7.8	27	9
32°C/8 hour adults	84	37.2	44	40.2	18.8	47	21
32°C/16 hour adults	84	18.0	21	21.0	5.9	28	7
"Natural" juveniles	64	16.3	26	19.3	7.7	39	12
32°C/8 hour juveniles	64	25.2	39	28.2	6.6	24	10
32°C/16 hour juveniles	64	11.8	18	14.8	9.8	66	15

period), we observe some slight enhancement of regenerative capacity following the second amputation (compare columns C and F, table 2). Which of these figures should be considered more significant will be discussed below.

DISCUSSION

The results confirm the variation in tail regenerative capacity observed in *A. carolinensis* by Maderson and Licht ('68). Although the incubator-maintained experimental groups were numerically small the value of V reported here for the first regeneration of nine adults was 25, very close to the V value of 24 for the 60 animals reported by Maderson and Licht. While our nine adults replaced an average of 35.0 mm of tail in 42 days compared with only 28.7 mm in the previous study

(see table 1), it is the variation, not the absolute figure, that is of interest here. Since all the animals studied in this series of experiments were collected within a known restricted geographical range, we suggest that variation is not primarily a function of population differences. The lack of consistent repeatability of regenerative performance for two consecutive regenerations suggests that no individual lizard is genetically a "good", "average" or "poor" regenerator. Variation *per se* apparently is not associated with either age group or with environmental conditions.

The present results permit a consideration of the effects of environment on tail regeneration and variation therein. An argument can be made that since lizards, like other ectotherms, live under

63

variable environmental conditions (notably with regard to temperature), maintaining them under constant conditions for experimental purposes may produce "abnormal" results. However, the present results and those reported by Maderson and Licht ('68) and Licht and Howe ('69) indicate that the amount of tail regeneration at 32°C constant temperature is maximal. The present data suggest that tail regeneration may be adversely affected by longer exposure to light *per diem*, not only absolutely, but also in terms of the amount of variation produced (assuming that the predictability of the results of regeneration is lessened as the process is impaired). If *A. carolinensis* is maintained under constant illumination, with sufficient thermal energy being available for regulatory behavior, little if any regenerative response is found (Maderson, unpublished). The "Natural" animals (Group A) reported on here shed their skins at approximately the same frequency as the incubator-maintained animals (Maderson et al., '70). Since both skin-shedding and tail regeneration are influenced in part by pituitary hormones (Licht and Howe, '69), it is of interest that varying temperature affects one growth phenomenon and not another. The apparent suppression of regenerative response in incubator-maintained animals under 16 hours illumination reported here warrants further investigation.

In evaluating the results of the second regeneration, we suggest that it should be scored as a percentage replacement of the original tail length rather than as a percentage replacement of the tail length prior to the second amputation, because we assume that the original tail length represented an adequate functional size for the individual concerned and that the "object" of a regenerative process should be to restore the optimal size. By these criteria the second regenerate is functionally worse than the first, since the net size of the tail is then even smaller and, therefore, further from the functionally adequate size. We tentatively suggest that the similarity in percentage replacement of removed tissue for both regenerations

indicates a developmental field phenomenon, and deserves further investigation.

This study and Maderson and Licht's ('68) indicate that the tail regeneration in *A. carolinensis* is highly variable; the cellular basis for this variability remains unknown. No comparable experimental data are available for other lacertilians and it is not known how important regenerative capacity is for *A. carolinensis* in the wild. Zweifel and Lowe ('66) drew attention to the fact that regenerated tails in *Xantusia vigilis* are both frequent and difficult to detect without close examination. They suggested that the ability for autotomy and subsequent regeneration is an important factor in the survival of this species in which individuals characteristically become sexually mature after four years, and have a low fecundity. Salthe and Maderson ('69) suggested that high coefficients of variability are associated with traits that are not important for the survival of individuals. The significance of tail regenerative capacities for the survival of various lizard species remains unknown.

ACKNOWLEDGMENTS

We wish to thank Misses R. Batachansky and J. Hoffman, and Messrs. J. Abramowitz and R. Fleigelman for technical assistance. Financial assistance was provided by N.I.H. grant CA-10844 (P.F.A.M.) and N.S.F. grants GB-5232 and GB-7749 (S.N.S.).

LITERATURE CITED

Lewontin, R. C. 1966 On the measurement of relative variability. Syst. Zool., *15:* 75-84.
Licht, P., and N. R. Howe 1969 Hormonal dependence of tail regeneration in the lizard *Anolis carolinensis.* J. Exp. Zool., *171:* 75-84.
Maderson, P. F. A., K. W. Chiu and J. G. Phillips 1970 Endocrine-epidermal relationships in squamate reptiles. Mem. Soc. Endocrinol., *18:* 259-284.
Maderson, P. F. A., and P. Licht 1968 Factors influencing rates of tail regeneration in the lizard *Anolis carolinensis.* Experientia, *24:* 1083-1086.
Salthe, S. N., and P. F. A. Maderson 1969 Physiological indeterminacy in *Anolis.* Am. Zool., *9:* 1069.
Zweifel, R. G., and C. H. Lowe 1966 The ecology of a population of *Xantusia vigilis,* the desert night lizard. Am. Mus. Nov., 2247: 1-57.

A Histological Study of the Regenerative Response in a Lizard, *Anolis carolinensis*

JOCELYN M. ZIKA

The limb of an urodele amphibian will not regenerate in the absence of sufficient innervation (Singer, '65; Goss, '69). It has been proposed that there is a direct relationship between regenerative ability and the quantity of nerve fibers (Singer, '47) or the total amount of axoplasm (Singer, Rzehak and Maier, '67) available at the amputation surface. This hypothesis has been tested experimentally by amputation and hyperinnervation of limbs in other animals, namely, frogs (Singer, '54) and lizards (Simpson, '61; Singer, '61; Kudokotsev, '62). The gross appearance of regenerated frog extremities reached the maximum development of a small two-digit limb. In lizards, a mounded adigital outgrowth of up to several millimeters in length was produced.

In none of the above cases was the histology of the early stages in the regenerative response described. Therefore it seemed of interest to compare postoperative changes in the amputated hyperinnervated limb of the lizard, *Anolis carolinensis*, with those of the amputated normally innervated limb. The postoperative histological changes in both the denervated and the normally innervated regenerating limbs of the newt, *Triturus viridescens*, have been described (Singer, '59). It is of particular interest that the changes reported here in both the hyper-

innervated and normally innervated limbs of *Anolis* approximate those of the denervated newt.

MATERIALS AND METHODS

Hindlimbs of 95 adult *Anolis carolinensis* were amputated through the proximal one-third of the thigh. Both skin and bone were further trimmed for a distance of about 1 mm from the amputation surface. Hyperinnervation was produced by deviation into an amputated stump of the contralateral sciatic nerve (Singer, '61). The animals were maintained in wire cages 35 cm square at a room temperature of $22 \pm 2°C$ and humidity of 40–45%. Mealworm larvae and water containing a liquid vitamin preparation (Homicebrin, Eli Lilly and Co.) constituted the animals' diet.

Both control (normally innervated) and experimental (hyperinnervated) limbs were fixed in groups of two or three at specific times after amputation: 6 and 12 hours, at daily intervals from 1 through 8 days, and then at 12, 17, 22, 30, 39, 48 and 56 days. In addition, 22 hyperinnervated animals, 15 of which showed a regenerative response, were sacrificed at various times from 13 weeks to 80 weeks.

Limbs were fixed in Bouin's solution, decalcified (Decal), imbedded in paraffin and serially sectioned at 10 μ. Tissues were stained for nerve fibers by the Bodian method and counterstained with an alcoholic solution of Orange G or with the Masson trichrome stain.

Nerve fiber counts were made on transversely sectioned hindlimbs according to the method of Singer (see Singer, '47; Zika and Singer, '65).

Postoperative tissue breakdown and reconstitution were essentially similar in both hyperinnervated and control limbs. Therefore the histological appearance of both groups will be described together. Following the initial postoperative changes some experimental limbs underwent more extensive tissue regeneration.

Wound closure and tissue repair

Gradual drying of the wound surface resulted in the formation within five days of a hard scab over the site. The scab remained in place until complete epithelization of the underlying wound surface had occurred. The area of the wound surface, following loss of the scab, appeared to be smaller than the original in both control and experimental limbs which did not produce a mound.

Epidermal migration over the wound surface began four days postoperatively, and was completed by 17 days in the controls. In the hyperinnervated limbs, complete epithelization was delayed until 39 days. The regenerating epidermis was multilayered, being thinnest at the advancing edge, and contained dividing cells. Tongues of regenerating epidermis were seen extending down into the wound tissues.

The appearance between 12 days (controls) and 17 days (experimentals) of fibrocellular connective tissue contiguous to the original wound margin was the first sign of dermal regeneration. Regrowth progressed slowly, and seemed to lag behind epidermal migration by several days. Inception of basement lamella formation below the regenerated epidermis nearest the old skin was observed 12 days after amputation. By 39 days it was reformed beneath the entire wound epidermis.

During the first several days after amputation, many blood cells were seen within the wound area. The leukocytes were primarily neutrophils which have been described as the phagocytes of early wound healing. In later stages, foci of a chronic inflammatory reaction were seen, consisting of fibroblasts and lymphocytes.

Cells which resembled the mesenchyme-like blastema cells of amphibians (Singer, '59; Salpeter, '65) were found scattered throughout the wound area in *Anolis* beginning at six days. However, many cells of the wound area were similar in appearance to fibroblasts, having a more oval nucleus and a generally elongated, flattened look. A subepidermal accumulation of cells which were primarily of the fibroblastic type appeared between 12 and 30 days in two animals of both control and experimental groups. At no time was a distal accumulation of solely mesenchyme-like cells, such as that reported for the amphibian regenerate, seen in these *Anolis* limbs.

In the present study regenerated blood capillaries were apparent only in peripheral regions five to six days after amputation. By 12 days vessels were present throughout the wound area and at 30 days the new dermis had been vascularized.

Sarcolysis of transected muscle began shortly after amputation and continued for several days. Sometimes the entire muscle fiber was broken down; in other cases sarcolysis was limited to the distal region. Muscle breakdown involved loss of cross-striations and apparent clumping of sarcoplasm which was later also destroyed. In a process similar to that described for amphibians (Thornton, '38) muscle nuclei surrounded by small amounts of cytoplasm were released into the wound area.

Muscle regeneration, which began about seven days after amputation, apparently occurred in two ways. Terminal protoplasmic buds containing many oval distal nuclei were seen continuous with the pre-existing transected muscle fibers. In addition, there were found narrow protoplasmic tubes containing longitudinally arranged central nuclei. These tubes did not seem to be continuous with pre-existing fibers, and probably were formed by the coalescence of so-called myoblasts as can occur in mammalian striated muscle regeneration (Karsner, '55). Cross-striations were detected in new muscle fibers between 12 and 17 days. Regenerating fibers extended a short distance into the wound area, and a few were attached to the proximal cartilage.

By six days, cells of the distal periosteum became widely separated, and the connective tissue fibers seemed scattered and reduced in number. Although at 12 days some cartilaginous matrix had been laid down adjacent to the bony shaft, dividing cells were still present. The cartilage formed an expanded sleeve and cap around the distal end of the bone by 30 days. Gradual ossification of the cartilage, beginning at 17 days, led to the formation of bony trabeculae extending partly or entirely across the width of the sleeve. Mitoses were seen in the cartilage cells as late as 30 days after amputation in the controls, whereas in the experimental limbs, division of cartilage cells persisted through 48 days.

Bone resorption, although apparently quite limited, was observed in early stages of bone repair. Large multinucleated osteoclasts were present in areas of bone resorption. These cells were seen near cavitations at the fractured end of the bone, as well as in association with eroded areas of the calcified cartilaginous matrix and newly formed bone. The channels which resulted from erosion of the calcified cartilage were filled with loose connective tissue (probably early bone marrow).

Following amputation, the transected nerves underwent the usual traumatic changes of axonal degeneration accompanied by the release and multiplication of Schwann cells in the distal part of the nerve trunk. Although regeneration of nerve fibers began on the second day after amputation, even by eight days invasion of the wound area was not especially intense in the controls. This situation contrasts with that in the newt, *Triturus,* in which after the first day or two the wound tissues are extensively invaded by nerves (Singer, '59). In *Anolis,* axial extension and branching into surrounding tissue progressed in subsequent days, but epidermal invasion was never extensive.

Morphology of mature outgrowths

In 15 of 25 hyperinnervated animals which were maintained for more than six weeks a darkly pigmented mound was seen protruding from the amputation surface between weeks five and seven. Growth of the mound was rapid for approximately 10 days. It then slowed and eventually halted after about 12 to 14 weeks of further growth. The final product was a somewhat conical mound ranging in length from 0.50 mm to 3.25 mm (table 1). In two cases, the mound was more or less flattened at the distal end, but there were no histological indications of hand formation. The 10 non-regenerating animals maintained a more or less flattened, healed-

TABLE 1

Regenerative response of hyperinnervated Anolis hindlimbs

Group	Animal	Length of mound	Adipose	Bone	Connective tissue	Muscle	Nerve
		mm					
I	117	0.5	0	0	++	0	++ [1]
	120	1.0	+	+++	+++	0	++
	122	0.5	0	0	++	0	+
	123	1.5	0	+++	++	0	++
	148	0.7	0	0	++	0	++
II	133	1.7	+++	0	++	0	+++
	134	3.2	+++	0	++	0	+++
	137	1.5	+++	0	++	0	+++
III	119	1.7	0	0	+++	+	++
	138	1.2	0	0	+++	+	++ +
IV	118	2.2	++	+	+++	+	++
	126	1.2	+	+++	+++	+	++
	136	2.0	0	++	+++	++	++
	139	3.0	0	+	+++	+	+++
	146	2.0	0	+	+++	+	++

[1] Scale ranges from 0, absent to +++, greatest amount seen.

67

over wound surface which was similar to that seen in the control animals.

Observations on all the mature mounds revealed limitations in their histogenic abilities (table 1). One group (I — five cases) was composed primarily of connective tissue, blood vessels and a large number of nerve fascicles, although in two mounds bone was also found (fig. 1). A second group (II — three cases) contained connective tissue, a limited number of blood vessels, adipose tissue in various amounts which was either concentrated distally or scattered throughout, and nerve fascicles (fig. 2). Connective tissue, nerve fibers, blood vessels and a small amount of proximal muscle were found (fig. 3) in a third group of mounds (III — two cases). The most histologically complete group of mounds (IV — five cases) was composed of connective tissue, some adipose tissue, nerve fibers, blood vessels, muscle and bone (fig. 4). The muscle was limited to the proximal area but in some cases bone extended nearly to the distal end of the mound.

The hyperinnervated limbs which showed no gross regenerative response were histologically similar to the non-regenerating control limbs. Some muscle had regenerated, a cap and sleeve of new bone surrounded and extended slightly above the transected end of the femur and a moderate amount of connective tissue filled the area between the epidermis and regenerated bone (fig. 5).

DISCUSSION

The developmental stages of regenerating newt limbs have been well described (Singer, '52; Hay and Fischman, '61). The most characteristic feature of this process is the accumulation of many large mesenchyme-like cells beneath the wound epidermis. These cells are termed blastema cells and are usually considered to give rise, at least partly, to the structures of the regenerate (Chalkley, '59; Rose and Rose, '65). A completely functional and histologically normal limb results from this process. It has also been shown that amputation of a denervated newt limb results in wound healing but no regeneration. Singer ('59) described the non-regenerat-

ing denervated newt limb as lacking or having a reduced number of mesenchymatous cells during the postamputational stages. The limited tissue regeneration which occurred was comprised of fibrocellular connective tissue, some muscle, and a cap and sleeve of cartilage around the transected end of the bone.

A cartilaginous mass surrounding the cut end of the bone and connective tissue formed the predominant tissues in healed amputated hindlimbs of *Lacerta vivipara* and *L. dugesii* (Bellairs and Bryant, '68) and forelimbs of *Anolis carolinensis* (Barber, '44). These authors reported the presence of a distal collection of undifferentiated blastemal or mesenchymal cells during early postoperative stages; however, no limb regeneration was observed. Cases from nature of abortive limb regeneration in lizards have been reported (Guyénot and Matthey, '28; Avel and Verrier, '30). Study of the internal composition of these structures revealed the presence of large amounts of cartilage in addition to lesser amounts of other limb tissues.

Amputated *Anolis* hindlimbs seen in this study resembled the histological picture drawn for amptuated denervated newt limbs. No cells which could definitely be classified as the amphibian type of blastemal cell accumulated subepidermally. In one case a conical mound was seen protruding from the amputation surface of a control limb at 48 days. Histological study revealed that a large mass of cartilage surrounding the transected end of the femur constituted the bulk of the conus which was not completely covered by a wound epithelium.

In 15 of 25 cases, supplementary innervation of the *Anolis* hindlimb gave rise to a mound which usually contained either cartilage or connective tissue as the predominant tissue. Simpson ('61) reported the presence of a cartilaginous model of the femur in addition to other tissues in *Lygosoma,* and Kudokotsev ('62) stated that some of the mounds formed in *Lacerta agilis* contained only cartilage and fibrous connective tissue whereas others contained muscle and nerves in addition to the cartilage and connective tissue. These reports of a regenerative response following hyper-

68

innervation of the lizard limb did not describe the early postoperative changes which preceded the appearance of the darkly pigmented conus. The results of the present study indicated qualitatively little difference between the morphological changes occurring after amputation of a normally innervated limb and one which was hyperinnervated immediately postoperatively. In the latter case, epithelization of the wound surface and thus formation of a complete skin seal were delayed, possibly allowing for more extensive proliferation of internal tissues at early postoperative stages. The regenerative response which followed hyperinnervation of the reptilian limb is therefore perhaps best described as extensive tissue repair, rather than the beginning of *de novo* regeneration of a functional replacement for the amputated structure. A similar idea was advanced by Skowran and Komala ('57) in their study of limb regeneration in an amphibian, *Xenopus laevis*.

Various traumatic influences may stimulate a regenerative response of the kind described in the hyperinnervated *Anolis* limbs. In another series of experiments, three out of 14 normal *Anolis* which were maintained under elevated postoperative humidity conditions for at least six weeks produced mounds which grossly and histologically similar to those seen after hyperinnervation (Zika, '68). Kudokotsev ('62) reported that in one limb out of four in which the deviated nerve was resected at the time of amputation, a mound composed of cartilage and fibrous connective tissue was formed. However, the traumatic influence of the operation itself probably is not the sole stimulus for the regenerative response. Singer ('61) and Simpson ('61) ran control series in which nerve deviation was performed and the nerve was then resected. No regenerative response was elicited. Also, the humidity experiment referred to above stimulated the formation of fewer mounds than did hyperinnervation.

Since the importance of the quantity of nerve fibers has been demonstrated in salamanders, it has been assumed that hyperinnervation of reptilian limbs would stimulate regeneration. Preliminary data obtained during this investigation (Zika, '68) indicated that in *Anolis*, deviation of the sciatic nerve may not raise the number of fibers in the hindlimb to a threshold level. Nerve counts on two animals indicated that the average nerve fiber number is 4.2 per $(100 \mu)^2$ of soft tissue at the amputation surface. Following deviation of the sciatic nerve the average number of fibers is calculated to be 6.2 per $(100 \mu)^2$, a number which is about 0.50 that of the *Anolis* forelimb (Zika and Singer, '65) and about 0.50 of the threshold quantity found to be necessary for limb regeneration in *Triturus* (Singer, '47). Thus it appears likely that, with the possible exception of Simpson's work ('61), the ability of hyperinnervation to induce lizard limb regeneration has not been adequately tested. A method for bringing larger amounts of nerve tissue into the limb must be developed before this procedure can be fully exploited.

LITERATURE CITED

Avel, M., and M.-L. Verrier 1930 Un cas de régénération hypotypique de la patte chez *La certa vivipara*. Bull. Biol. Fr. et Belg., 64: 198–203.

Barber, L. W. 1944 Correlations between wound healing and regeneration in fore-limbs and tails of lizards. Anat. Rec., 89: 441–454.

Bellairs, A. d'A., and S. V. Bryant 1968 Effects of amputation of limbs and digits of Lacertid lizards. Anat. Rec., 161: 489–496.

Chalkley, D. T. 1959 The cellular basis of limb regeneration. In: Regeneration in Vertebrates. C. S. Thornton, ed. University of Chicago Press, Chicago, pp. 34–58.

Goss, R. J. 1969 Principles of Regeneration. Academic Press, New York and London.

Guyénot, E., and R. Matthey 1928 Les processus régénératifs dans la patte postérieure du lézard. Arch. Entwicklungsmech., 113: 520–529.

Hay, E. D., and D. A. Fischman 1961 Origin of the blastema in regenerating limbs of the newt *Triturus viridescens*. Develop. Biol., 3: 26–59.

Karsner, H. T. 1955 Human Pathology. J. B. Lippincott Co., Philadelphia and Montreal.

Kudokotsev, V. P. 1962 Stimulation of the regeneration process in the extremities of lizards by the method of supplementary innervation. Doklady Akad. Nauk SSSR, Biol. Sci. Sect., 142: 101–104.

Rose, F. C., and S. M. Rose 1965 The role of normal epidermis in recovery of regenerative ability in x-rayed limbs of *Triturus*. Growth, 29: 361–393.

Salpeter, M. M. 1965 Disposition of nerve fibers in the regenerating limb of the adult newt, *Triturus*. J. Morph., *117*: 201–211.

Simpson, S. B., Jr. 1961 Induction of limb regeneration in the lizard, *Lygosoma laterale*, by augmentation of the nerve supply. Proc. Soc. Exp. Biol. Med., *107*: 108–111.

Singer, M. 1947 The nervous system and regeneration of the forelimb of adult *Triturus*. VII. The relation between number of nerve fibers and surface area of amputation. J. Exp. Zool., *104*: 251–265.

———— 1952 The influence of the nerve in regeneration of the amphibian extremity. Quart. Rev. Biol., 27: 169–200.

———— 1954 Induction of regeneration of the forelimb of the postmetamorphic frog by augmentation of the nerve supply. J. Exp. Zool., *126*: 419–471.

———— 1959 The influence of nerves on regeneration. In: Regeneration in Vertebrates. C. S. Thornton, ed. University of Chicago Press, Chicago, pp. 59–80.

———— 1961 Induction of regeneration of body parts in the lizard, *Anolis*. Proc. Soc. Exp. Biol. Med., *107*: 106–108.

———— 1965 A theory of the trophic nervous control of amphibian limb regeneration, including a re-evaluation of quantitative nerve requirements. In: Regeneration in Animals and Related Problems. V. Kiortsis and H. Trampusch, eds. North-Holland Publishing Co., Amsterdam, pp. 20–32.

Singer, M., K. Rzehak and C. Maier 1967 The relation between the caliber of the axon and the trophic activity of nerves in limb regeneration. J. Exp. Zool., *166*: 89–98.

Skowron, S., and Z. Komala 1957 Regeneration of limbs of *Xenopus laevis* after metamorphosis. Fol. Biol., 5: 53–72.

Thornton, C. S. 1938 The histogenesis of muscle in the regenerating forelimb of larval *Amblystoma punctatum*. J. Morph., 62: 17–47.

Zika, J. 1968 A histological study of limb regeneration in the lizard, *Anolis carolinensis*. Ph.D. thesis.

Zika, J., and M. Singer 1965 The relation between nerve fiber number and limb regenerative capacity in the lizard, *Anolis*. Anat. Rec., *152*: 137–140.

PLATE

PLATE 1

EXPLANATION OF FIGURES

A, adipose; CT, connective tissue; M, muscle; N, nerve.

1 An amputated hyperinnervated stump on which a group I growth, containing connective tissue and nerve has formed. Arrows indicate the approximate amputation level. $\times$ 78.

2 An amputated hyperinnervated stump on which a group II growth containing nerve and adipose tissue has formed. This growth was located over the area of the wound surface in which the innervation was concentrated. $\times$ 78.

3 A group III growth formed on an amputated hyperinnervated stump. Long arrows indicate the approximate amputation level. A short arrow indicates the area of the growth in which a limited amount of muscle was found. $\times$ 78.

4 Group IV growth formed after amputation of a hyperinnervated limb. Regenerated bone is located between the muscle and nerve. $\times$ 78.

5 Non-regenerating hyperinnervated limb: note limited amount of new bone (clear arrow) and subcutaneous connective tissue. Approximate level of amputation is indicated by solid arrows. $\times$ 78.

PLATE 1

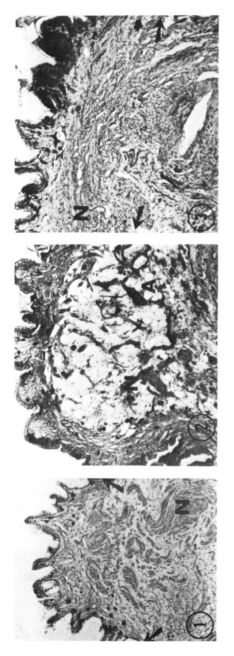

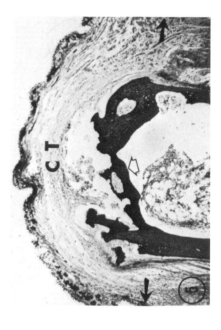

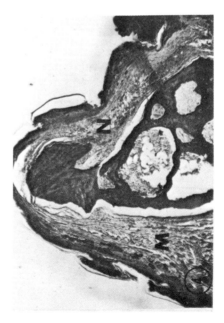

Cytology of Limb Regeneration in Amphibians

The Urodele Limb Regeneration Blastema: A Self-Organizing System

I. Differentiation *in Vitro*[1]

DAVID L. STOCUM[2]

INTRODUCTION

Regeneration of an amputated urodele limb occurs by the formation of a blastema from which the missing limb components are subsequently reformed. The cells of the blastema originate by a limited morphological dedifferentiation of most of the mesodermally derived stump tissues local to the amputation surface (Butler and O'Brien, 1942; Chalkley, 1954; Hay and Fischman, 1961). The dedifferentiated cells divide rapidly, resulting in the elongation of the blastema into a conical configuration. Histologically, the cone is composed of a cytologically homogeneous, overtly undifferentiated population of mesenchyme-like cells. From these cells, all of the missing limb parts are subsequently redifferentiated in continuity with the stump tissues to produce an anatomically and functionally complete limb.

One of the long-standing mysteries of amphibian regeneration involves the question of what factors control the morphogenesis and differentiation of new limb parts from the undifferentiated blastema.

[1] Part of a dissertation presented to the faculty of the Graduate School of Arts and Sciences of the University of Pennsylvania in partial fulfillment of the requirements for the degree of Doctor of Philosophy, 1968. Research supported by U.S.P.H.S. DE-02047, a program project grant under the direction of Dr. C. E. Wilde, Jr.

[2] Predoctoral Fellow, U.S.P.H.S. Developmental Biology Training Grant 5 T1-GM-849-05 and U.S.P.H.S. T0 1 DE-00001-11.

The blastema never forms more than those structures that are distal to the plane of amputation, and the regenerated elements are harmoniously integrated into the tissues of the stump. One explanation that has long been in vogue to account for this harmonious development holds that the differentiated tissues of the limb stump act to induce the undifferentiated blastema to redifferentiate lost parts, in a manner analogous to induction phenomena in the embryo. This concept has been based on the rather general experience that undifferentiated blastemas transplanted to foreign sites usually fail to differentiate, whereas blastemas that have begun differentiation prior to transplantation are able to do so (Weiss, 1930; see Goss, 1961 for review of this subject). In the latter case, the stump has presumably transmitted inductive messages to the blastema by the time transplantation takes place. These messages are usually envisaged as being chemical in nature (J. Needham, 1942; Faber, 1965).

The cone stage blastema is, for the most part, dependent on a trophic stimulus from the limb nerves for its continued development and will often resorb if the limb is denervated (Schotté and Butler, 1944; Butler and Schotté, 1949; Singer, 1952). The trophic effect is quantitative in the sense that a threshold quantity of neuronal substance is required to maintain the regenerative process (Rzehak and Singer, 1966). Blastemas transplanted to foreign sites are temporarily disconnected from a nerve supply, and their failure to undergo development has usually been associated with graft resorption (see Goss, 1961). However, several investigators have obtained partial differentiation and morphogenesis in transplanted conical blastemas that were not totally resorbed (Mettetal, 1952; Pietsch, 1961; Faber, 1965). These results indicate that at this stage the blastema may actually be considerably independent of the stump for its subsequent differentiation and morphogenesis.

Another technique for testing the ability of a developing system to undergo independent development is to isolate the system *in vitro*. Loss of cells by resorption into an animal body is precluded by this method. Attempts have been made to obtain organogenesis and differentiation of urodele limb blastemas *in vitro*, but have proved unsuccessful (LeCamp, 1947, 1948; Fimian, 1959), although the embryonic urodele limb bud is capable of organogenesis and differentiation *in vitro* (Wilde, 1950). The present report is concerned with experiments designed to test the self-differentiation capacity of the urodele limb blastema by utilizing a tissue culture method developed for this purpose. The results indicate that the limb stump has no inductive influence on the differentiation of the blastema once the latter has reached the cone stage.

MATERIALS AND METHODS

All experiments were carried out on the larvae of *Ambystoma maculatum* (Shaw), obtained as eggs from Dr. Glenn Gentry, Donelson, Tennessee, or from ponds in the vicinity of Princeton, New Jersey. The animals were reared in spring water at 21°C in the laboratory, and were fed *Tubifex* every 4 days. The water was changed at each feeding. Regeneration blastemas were obtained by anesthetizing the larvae in 1:1000 MS:222 (Sandoz) in spring water and amputating their forelimbs bilaterally, just proximal to the elbow.

Culture Groups

Three stages of regeneration were employed for culture, (a) the histologically undifferentiated cone, (b) the palette, in which precartilage condensations have appeared, and (c) the notch, in which the first two digits have become externally visible. The capacity for autonomous development of these stages was tested by culturing blastemas of each stage in the absence of the limb stump. In order to determine whether the presence of stump would affect the development of the regenerate, blastemas of each stage were also cultured with a short segment of limb stump. Thus, there were six groups of cultures prepared. A total of 103 cultures was distributed among these six groups.

Culture Medium

Several media currently in use for amphibian tissue culture were screened to find one which was suitable for blastema culture. None of these media proved satisfactory (see Results); therefore, a new medium was designed for this purpose. This medium consisted of 88 parts of Leibovitz L-15 defined medium (Grand Island Biological Co.) diluted to 80% with distilled water, 10 parts of fetal calf serum (Microbiological Associates), and 2 parts of beef embryo extract (Microbiological Associates). No antibiotics were added. Average initial pH of the medium was 7.4 and the average osmolality, as measured on a Fiske Osmometer Model G-62 (Fiske Associates) was 263 milliosmoles.

The medium required no gassing, since its buffer system is composed predominantly of free base amino acids in the L-15 component, rather than carbonate-bicarbonate (see Leibovitz, 1963).

Culture Procedures

Aseptic technique was used throughout the culture procedures. An initial attempt was made to culture regenerates within their epithelium, but it was found impossible to decontaminate the latter with antibiotic solutions or astringent agents with any consistent degree of success. In addition, in those cases where decontamination was successful, it appeared that the healed epidermis was impermeable to the passage of nutrients from the culture medium. The results, therefore, are based only on the 103 epidermis-free cultures prepared by the following methods.

Stumpless regenerates. Forelimbs with their regenerates were removed from the animals and freed of epidermis by treatment for 45 minutes in calcium-magnesium-free Holtfreter solution containing 0.1% EDTA at room temperature (about 25°C). After transfer to fresh Holtfreter solution, the blastemas were carefully separated from the stump with fine iris scissors or knives (Heiss, Germany). They were then washed in three changes of fresh Holtfreter solution and in two changes of culture medium. This procedure kept contamination by microorganisms to a minimum. Operational controls were employed to check the accuracy of the operation in excluding stump tissues. These consisted of several blastemas which were routinely removed from the second medium wash of each batch of cultures and immediately fixed. In some cases, the controls were fixed in Bouin's solution, sectioned at 10 μ and stained with iron hematoxylin and light green. In other cases, they were fixed in Gregg's solution and stained for cartilage with methylene blue by the Van Wijhe method, as modified by Gregg and Butler (published in Hamburger, 1960).

Each blastema was explanted onto a coverslip in a small, sealable culture dish (internal diameter, 30 mm; height, 12 mm, Bellco Glass Co.) and just covered with medium. All cultures were allowed to develop at room temperature for 25 days. It was unnecessary to feed the cultures over the duration of the culture period.

Regenerates cultured with stump. The procedures for preparing these cultures were exactly the same as those for stumpless explants, with the exception that a segment of stump about equal to the length of the blastema was carried along in the explant.

Histological Preparations

Cultures were terminated at 25 days by washing them on their coverslips in Holtfreter solution, followed by fixation. Individual cultures in each of the six culture groups were treated in either of two ways: (a) fixed on the coverslip in Gregg's solution and stained *in toto* for cartilage with methylene blue as described above, or (b) fixed in Bouin's solution, followed by removal of the explant from the coverslip, leaving the monolayer attached. The latter explants were sectioned at 10 μ, and along with their monolayers on the coverslips, stained with iron hematoxylin and light green, or with toluidine blue. All sectioned explants were examined for muscle, in addition to cartilage.

RESULTS

An important condition for the study of any developing system *in vitro* is selection of a culture medium which provides a milieu in which the developmental potential of the system may be expressed. Media currently in use for amphibian tissue culture, including Holtfreter saline, were able to support the survival of blastema cells, but did not promote their organogenesis or differentiation. Gordon and Wilde (1965) found that differentiation of embryonic chick heart cells *in vitro* depended to a great extent on adjusting the osmolality of the culture medium to an optimal level, which would presumably approximate that of avian intercellular fluid. The osmolality of salamander body fluid, as calculated from its freezing-point depression (given in Heilbrunn, 1952) is 258 milliosmoles. Measurements of osmolality on media which merely supported blastema cell survival revealed that their osmolalities ranged from 20 to 60 milliosmoles below this value. Therefore, the final medium employed in the present study was designed to have an osmolality approximating that of salamander body fluid. Blastemas explanted to this medium readily underwent differentiation.

Development of Stumpless Explants

The state of morphogenesis and differentiation of operational control regenerates is illustrated in Figs. 1–3. The cone stage blastema (Fig.

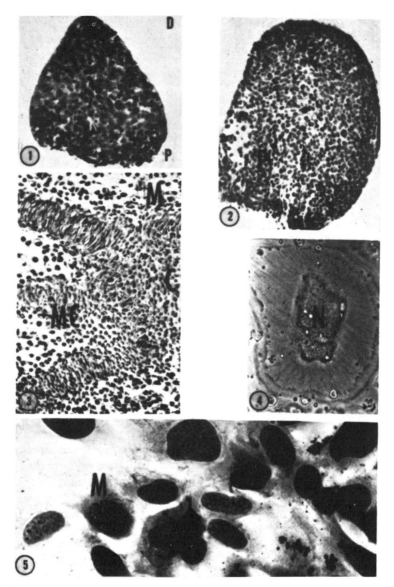

FIG. 1. Longitudinal section through an isolated cone stage blastema. The cells of the cone are cytologically homogeneous and overtly undifferentiated. D, distal; P, proximal. × 65.

81

1) is composed of an unorganized population of cytologically homogeneous cells. The cone does not stain for cartilage with methylene blue. By the palette stage (Fig. 2) redifferentiation of the blastema has begun. The blastema is flattened along its dorsoventral axis and precartilage condensations have appeared in its proximal portion. The reaction of these condensations to methylene blue ranges from none to light; very little matrix has been produced by the chondrocytes at this time. The notch stage is distinguished externally by the appearance of the first two digits. Figure 3 illustrates a section of a relatively advanced notch stage regenerate. Most of the cartilage anlagen are laid down, and myoblasts are aligned parallel to their long axes. Striations do not appear in the muscle cells until sometime later, however. The proximal cartilage anlagen now stain lightly to moderately with methylene blue. Examination of several such controls indicated that stump tissues were completely excluded by the surgical procedures employed in isolating the blastema.

In the living state, cultures of all stages exhibited two morphological phases, one a sheet of cells which migrated from the explant on all sides, the other a central mass of cells which underwent progressive internal organization with time. Numerous mononucleate circular and giant cells began migrating from cone and palette explants within 24 hours and from notch explants within 3 days.

The giant cells often increased in size with time, but the mechanism of this size increase is unknown. These cells were characterized by their large clusters of highly basophilic nuclei and by the presence of a band of granular material around the circumference of their peripheral cytoplasm (Fig. 4). The initial outwandering of giant cells was shortly followed by migration of a sheet of spindle- or stellate-shaped cells which became predominant in cone and palette cultures by day 8

FIG. 2. Longitudinal section through an isolated palette stage blastema. Note the precartilage condensations (P) in the proximal portion of the blastema. × 65.

FIG. 3. Longitudinal section through the carpal (C) and metacarpal (MC) regions of a relatively advanced notch stage blastema. Myoblasts (M) are aligned parallel to the cartilage anlagen. × 65.

FIG. 4. Living multinucleate giant cell of a cone stage explant. N, nuclear cluster. Phase contrast. × 163.

FIG. 5. Portion of the monolayer of a palette stage explant. M, mitotic prophase. × 260.

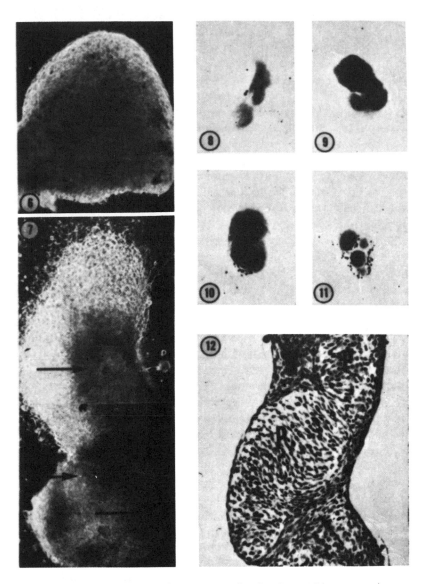

FIG. 6. Living cone stage blastema immediately after explantation. No internal organization is evident. Phase contrast. × 39.

FIG. 7. Same explant 24 days later. Arrows indicate organized regions that developed during the culture period. M, monolayer. Phase contrast. × 39.

and in notch cultures by day 12. Beginning at about day 14 of culture, frequent divisions were observed in the cells of this sheet in cultures of all stages. Figure 5 illustrates a portion of a typical monolayer with a dividing cell. No divisions were ever noted in multinucleate giant cells.

The increasing organization of the central, compact cell mass of an explant is illustrated by the cone stage regenerate depicted in Figs. 6 and 7. Figure 6 represents the blastema immediately after explantation. By 8 days discrete areas of organization became apparent within the regenerate, and by 24 days its final developmental status had been attained (Fig. 7). This culture underwent an apparent increase in size during 25 days in vitro (Figs. 6 and 7 are at the same magnification). Increase in size over the period of culture was typical of explanted blastemas.

Whole mounts of explants of all stages stained heavily with methylene blue in discrete regions after 25 days in vitro, indicating that differentiation of skeletal elements had taken place. Figures 8 and 9 represent whole mounts of cone stage cultures. Both explants differentiated three elements, two of which were rod-shaped and fused along their longitudinal axes. Figures 10 and 11 represent palette and notch stage explants, respectively. The palette differentiated nodular and rodlike skeletal elements. Four separate elements differentiated from the notch explant which appeared to have carpal-like morphology. Cultures of all regenerate stages stained much more heavily with methylene blue than did their corresponding operational controls.

In sectioned material, the organized regions revealed by methylene blue appeared to be precartilages composed of tiers or whorls of widely spaced cells surrounded by a refractile matrix (Fig. 12). The matrix stained metachromatically with toluidine blue and varying amounts of fibrous material exhibiting metachromasia were present in many explants. Sections also revealed the presence of immature or well-differentiated striated muscle in cultured blastemas of all stages. Figure 13 represents a section of a notch stage explant which differenti-

FIGS. 8–11. Twenty-five-day methylene blue-stained cone (Figs. 8 and 9), palette (Fig. 10), and notch (Fig. 11) stage explants exhibiting cartilage differentiation. Explant outlines and cell monolayers do not take the stain. × 16.

FIG. 12. Longitudinal section through the cone stage explant of Fig. 7. Note the organized regions (P) of precartilage. A possible developing joint is indicated by the arrow. × 65.

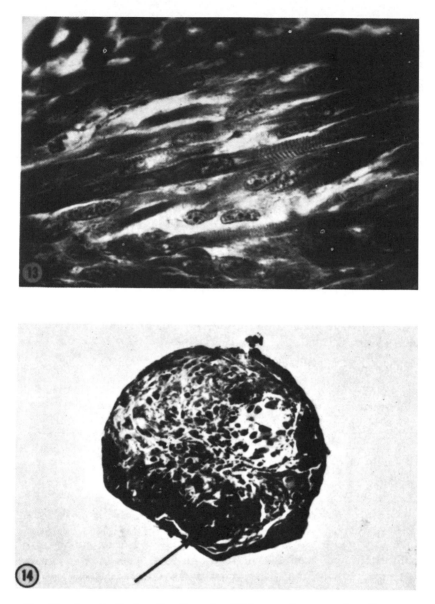

Fig. 13.   Longitudinal section through a 25-day notch stage explant illustrating striated muscle. Large amounts of muscle were present throughout the explant. × 312.

ated a large quantity of muscle, and illustrates the extent to which muscle may be formed by the regenerate *in vitro*. Individual notch stage explants differentiated muscle in greater quantities than did individual explants of cone or palette stages.

Table 1 summarizes the percentage of cases in each culture group which exhibited distinct differentiation. Nearly 64% of cone stage blastemas differentiated precartilage where none was present at the time of explantation. A higher frequency of precartilage differentiation was observed in palette and notch cultures. However, precartilage

TABLE 1

DIFFERENTIATION OF PRECARTILAGE AND STRIATED MUSCLE IN 103
REGENERATES CULTURED FOR 25 DAYS *in Vitro* WITH
(+) AND WITHOUT (−) STUMP

Culture groups	Cone		Palette		Notch	
	(−)	(+)	(−)	(+)	(−)	(+)
No. of cases	22	13	26	10	21	11
No. of cases examined for precartilage	22	13	26	10	21	11
% Differentiating precartilage	63.6	7.7	73.1	60.0	85.7	54.6
No. of cases examined for muscle	13	11	20	7	19	8
% Differentiating muscle	7.7	0.0	45.0	0.0	73.7	50.0
% Regenerates disappearing	0.0	77.7	0.0	20.0	0.0	9.1

condensations were present in palette and notch stages prior to culture. Although methylene blue-stained whole mounts indicated that an increase in the amount of skeletal material occurred in these explants, sections showed that the differentiative maturity of this material did not advance beyond the status of precartilage. Thus precartilage formation probably represents an advance in differentiation only in cone stage explants. Muscle, on the other hand, had differentiated to the mature (striated) condition in nearly all cases where it was found. As seen from Table 1, the percentage of cases differentiating this tissue

FIG. 14. Section through a 25-day cone stage blastema cultured with stump. The blastema has disappeared, leaving only the stump as a morphological entity. The stump itself has become transformed into an amorphous mass of tissue containing large amounts of heavy fibrous material (arrow). × 78.

increased rapidly with the advance in stage of the regenerate at the time of explantation. Muscle was not present in either cone or palette stage operational controls at the beginning of culture, but was present in the form of myoblasts in notch stage controls. It is evident that the ability to differentiate muscle became considerably greater in the latter two stages than in the cone stage.

In spite of the differences in maturity of differentiation of skeletal and muscular elements, it is clear that the state of differentiation of explanted stumpless regenerates of all stages became, in one form or another, far advanced over that present at the time of explantation.

Development of Regenerates Cultured with Stump

Regenerates cultured with stump exhibited marked contrasts to those cultured without stump. Migration of the sheet of spindle- and stellate-shaped cells so characteristic of stumpless explants did not occur in these cultures, and their cellular monolayer remained composed of circular and giant cells throughout the period of culture. Folding of the regenerate so that it came to lie parallel to the stump was a common occurrence. A striking feature of cultures of this type was that the blastemal portion of cone explants completely disappeared in 77.7% of the cases, leaving only the stump as a morphological entity, whereas in palette and notch explants this portion rarely disappeared (Table 1). Figure 14 illustrates this condition. The stump is all that remains of the explant; there is no sign of a blastema. Those cone blastemas which did survive rarely formed precartilage and never differentiated muscle. The frequency of precartilage and muscle differentiation in palette and notch explants was greatly reduced in comparison to that of stumpless explants (Table 1). Moreover, the stump segment itself often appeared to have been transformed into an amorphous mass of tissue containing large amounts of fibrous material that stained metachromatically with toluidine blue (see Fig. 14). It is evident that inclusion of stump did not enhance the differentiation of an explanted regenerate at any stage. If anything, the stump appeared to be detrimental to the development of the blastema.

DISCUSSION

The results of the present culture experiments support the position that the young limb regeneration blastema is capable of self-differentiation in the absence of the limb stump. The histologically

87

undifferentiated cone stage blastema was able to differentiate distinct precartilage and mature striated muscle *in vitro,* although the latter tissue was infrequently encountered at this stage. Mature muscle was differentiated at high frequencies in cultures of older regenerate stages, but skeletal differentiation remained at the precartilaginous level of operational controls. It would appear that the *in vitro* conditions employed in the present study for blastema culture are more favorable to the development of muscle than cartilage.

It might be expected that, if the limb stump exerts an inductive influence on the blastema, the latter would exhibit more extensive differentiation than that observed for stumpless blastemas. However, this does not appear to be the case. Inclusion of stump in an explant of any stage did not enhance, and even appeared detrimental to, development of the regenerate. Why the stump should inhibit blastema differentiation *in vitro* is of some interest, but the data offer no ready explanation for this phenomenon.

In the present study, cultured regenerates were able to differentiate only in a medium whose osmolality had been adjusted to approximate that of salamander body fluid. Previous attempts to obtain differentiation of the urodele limb blastema *in vitro* have failed (LeCamp, 1947, 1948; Fimian, 1959). It is possible that the osmolalities of the media employed in the latter studies were unfavorable to differentiation of the explants. In this connection, Simpson and Cox (1967) have demonstrated that, by changing the composition of the culture medium, they could promote or suppress the differentiation of striated muscle in cultures of dissociated myoblasts of lizard (*Anolis*) tail blastemas.

In view of the well-known dependence of the early blastema on innervation (Schotté and Butler, 1944; Butler and Schotté, 1949; Singer, 1952), it is surprising that cone stage blastemas were able to undergo differentiation *in vitro* with any great frequency, since they remained denervated over the duration of the culture period. Yntema (1959) has shown that limbs of urodele larvae which have been raised under aneurogenic conditions are completely independent of innervation for their regeneration. It is possible that the cells of an aneurogenic limb blastema or a blastema explant may themselves produce a trophic factor the same as, or similar to, the one supplied by the nerves (see Singer, 1965). If produced in large enough quantity, it is conceivable that a "conditioning" effect may be mediated by the factor in the explant medium.

Although recognizable precartilage and muscle developed in cul·
tured regenerates, these tissues were rarely arranged in normal limb
morphology. This can be attributed in large measure to the fact that
the explants were not enclosed in their epithelia and thus were able
to spread histiotypically, a phenomenon which will obviously alter the
character of any expected morphological configuration. On the other
hand, clear limb organogenesis has been demonstrated in explants of
embryonic urodele limb buds encased in their epithelia (Wilde, 1950).
Likewise, Hauser (1965) obtained morphologically normal regenerates
from segments of larval *Xenopus* tails cultured with their epidermis in
Holtfreter saline. Notochord was redifferentiated from the blastemas
formed on the tail explants, but muscle remained at the stage of myoblasts. It would be interesting to determine whether the differentiation
and morphogenesis of the isolated blastema would be more normal
if the latter were allowed to develop within its epithelium and under
conditions in which innervation would be restored. Such considerations will be the subject of a subsequent paper.

In conclusion, the results obtained in the present study suggest that
differentiative information specifying the formation of new cartilage
and muscle is programmed within the cells of the blastema by the time
it has arrived at the cone stage. Thus, even at this stage the blastema
is not an indifferent structure whose differentiation depends on inductive messages from the stump, but is a self-differentiating system.

SUMMARY

It has long been held that the overtly undifferentiated conical regeneration blastema is incapable of differentiation in the absence of inductive influences from the differentiated tissues of the stump. The validity
of this concept was tested by culturing cone, palette, and notch stage
limb blastemas of larval *Ambystoma maculatum in vitro*, both with
and without stump. Blastemas of all three stages were able to differentiate precartilage and striated muscle *in vitro* in complete absence
of the stump. Inclusion of stump in an explant did not enhance the
frequency of differentiation of precartilage and muscle in the blastema,
and actually appeared to be detrimental to development of the regenerate. It has been concluded that by the time the blastema has reached
the cone stage, it is not dependent upon inductive messages from the
stump for differentiation, but is a self-differentiating system.

The author wishes to express his appreciation to Dr. Charles E. Wilde, Jr., for his encouragement and advice throughout the course of this investigation, and for critical reading of the manuscript.

REFERENCES

BUTLER, E. G., and O'BRIEN, J. P. (1942). Effects of localized X-irradiation on regeneration of the urodele limb. *Anat. Record* 84, 407–413.

BUTLER, E. G., and SCHOTTÉ, O. E. (1949). Effects of delayed denervation on regenerative capacity in limbs of urodele larvae. *J. Exptl. Zool.* 112, 361–392.

CHALKLEY, D. T. (1954). A quantitative histological analysis of forelimb regeneration in *Triturus viridescens*. *J. Morphol.* 94, 21–71.

FABER, J. (1965). Autonomous morphogenetic activities of the amphibian regeneration blastema. *In* "Regeneration in Animals" (V. Kiortsis and H. A. L. Trampusch, eds.), pp. 404–418. North Holland Publ., Amsterdam.

FIMIAN, W. J. (1959). The *in vitro* cultivation of amphibian blastema tissue. *J. Exptl. Zool.* 140, 125–144.

GORDON, H. P., and WILDE, C. E., JR. (1965). "Conditioned" medium and heart muscle differentiation: contrast between explants and disaggregated cells in chemically defined medium. *Exptl. Cell Res.* 40, 438–442.

GOSS, R. J. (1961). Regeneration of vertebrate appendages. *In* "Advances in Morphogenesis" (M. Abercrombie and J. Brachet, eds.), pp. 103–152. Academic Press, New York.

HAMBURGER, V. (1960). "A Manual of Experimental Embryology," revised ed., p. 196. Univ. of Chicago Press, Chicago, Illinois.

HAUSER, R. (1965). Autonome Regenerationsleistungen des larvalen Schwanzes von *Xenopus laevis* und ihre Abhängigkeit vom Zentralnervensystem. *Roux' Arch. Entwicklungsmech. Organ.* 156, 404–448.

HAY, E. D., and FISCHMAN, D. A. (1961). Origin of the blastema in regenerating limbs of the newt *Triturus viridescens*. An autoradiographic study using tritiated thymidine to follow cell proliferation and migration. *Develop. Biol.* 3, 26–59.

HEILBRUNN, L. V. (1952). "An Outline of General Physiology," 3rd. ed., p. 134. Saunders, Philadelphia, Pennsylvania.

LECAMP, M. (1947). Les tissus de régénération en culture *in vitro*. *Compt. Rend. Acad. Sci.* 224, 674.

LECAMP, M. (1948). Régénération chez le Triton *in vitro* et *in vivo*. *Compt. Rend. Acad. Sci.* 226, 695.

LEIBOVITZ, A. (1963). The growth and maintenance of tissue-cell cultures in free gas exchange with the atmosphere. *Am. J. Hyg.* 78, 173–180.

METTETAL, C. (1952). Action du support sur la differenciation des segments proximaux dans les régénérats de membre chez les Amphibiens Urodèles. *Compt. Rend. Acad. Sci.* 234, 675–676.

NEEDHAM, J. (1942). "Biochemistry and Morphogenesis," pp. 430–442. Cambridge Univ. Press, London and New York.

PIETSCH, P. (1961). Differentiation in regeneration. I. The development of muscle and cartilage following deplantation of regenerating limb blastemata of *Amblystoma* larvae. *Develop. Biol.* 3, 255–264.

RZEHAK, K., and SINGER, M. (1966). Limb regeneration and nerve fiber number in *Rana sylvatica* and *Xenopus laevis*. *J. Exptl. Zool.* 162, 15–22.

SCHOTTÉ, O. E., and BUTLER, E. G. (1944). Phases in regeneration and their dependence on the nervous system. *J. Exptl. Zool.* 97, 95–122.

SIMPSON, S. B., and COX, P. G. (1967). Vertebrate regeneration system: culture *in vitro*. *Science* 157, 1330.

SINGER, M. (1952). The influence of the nerve in regeneration of the amphibian extremity. *Quart. Rev. Biol.* 27, 169–200.

SINGER, M. (1965). A theory of the trophic nervous control of amphibian limb regeneration, including a re-evaluation of quantitative nerve requirements. *In* "Regeneration in Animals" (V. Kiortsis and H. A. L. Trampusch, eds.), pp. 20–32. North Holland Publ., Amsterdam.

WEISS, P. (1930). Potenzprüfung am Regenerationsblastem. II. Das Verhalten des Schwanzblastems nach Transplantation an die Stelle der Vorderextremität bei Eidechsen (Lacerta). *Roux' Arch. Entwicklungsmech. Organ.* 122, 379–394.

WILDE, C. E., JR. (1950). Studies on the organogenesis *in vitro* of the urodele limb bud. *J. Morphol.* 86, 73–113.

YNTEMA, C. L. (1959). Regeneration in sparsely innervated and aneurogenic forelimbs of *Amblystoma* larvae. *J. Exptl. Zool.* 140, 101–124.

The Urodele Limb Regeneration Blastema:
A Self-Organizing System

II. Morphogenesis and Differentiation of Autografted Whole and Fractional Blastemas[1]

DAVID L. STOCUM[2]

INTRODUCTION

Regeneration blastemas of larval urodele limbs fail to undergo normal morphogenesis when cultured *in vitro* without their epidermis (Stocum, 1968b), but will develop more normally if transplanted within their epithelium to another site on the animal body. However, the extent to which a grafted blastema is able to develop seems to depend on the presence or absence of limb stump in the graft. In the absence of stump, histologically undifferentiated cone stage blastema transplants usually form only hand structures, whereas essentially complete limbs are formed if a segment of stump is included in the graft (David, 1932; Mettetal, 1939; Pietsch, 1961). Indeed, Faber (1960, 1965) has presented evidence that the morphogenesis of the hand in a regenerating limb is autonomous, while that of more proximal limb structures is dependent on stump induction, a hypothesis

[1] Part of a dissertation presented to the faculty of the Graduate School of Arts and Sciences of the University of Pennsylvania in partial fulfillment of the requirements for the degree of Doctor of Philosophy, 1968. Research supported by U.S.P.H.S. DE-02047, a program project grant under the direction of Dr. C. E. Wilde, Jr.

[2] Predoctoral Fellow, U.S.P.H.S. Developmental Biology Training Grant 5 T1-GM-849-05 and U.S.P.H.S. To 1 DE-00001-11.

first suggested by Mettetal in 1952. The examination of this hypothesis is the central concern of the present report.

Faber's hypothesis is based on the following experiments. By tracing the fate of carbon marks inserted into the tips of cone stage blastemas obtained by amputation through the humerus of axolotl larvae, Faber found that the material destined to form the hand arose by proliferation of cells distal to the mark. The marked mesenchyme formed the upper and lower arm of the regenerate. However, a cone blastema marked in the same way formed only hand structures ("distalized") when grafted to the back. In most of these cases, the carbon marks were pulled to the base of the transplant by resorption of cells. But in a few cases, carbon was found in the hand itself, leading to the conclusion that, at the cone stage, cells that normally would form upper and lower arm actually possess differentiation tendencies which are markedly more distal.

In a second experiment, Faber transplanted the distal and proximal halves of palette stage blastemas separately. The proximal halves were provided with a distal carbon mark. The distal half transplants, which contained the rudiments of carpals and digits, formed just those structures. The proximal half transplants, which contained the rudiments of upper and lower arm skeleton, were covered with fresh wound epithelium and dedifferentiated. Subsequently, most of these grafts distalized and formed carpals and digits also, although a few differentiated lower arm skeleton in addition. In those cases which formed the lower arm, the carbon marks were found within this region, as expected. In the majority of those grafts which formed only carpals and digits, the carbon particles were pulled to the base of the transplant by resorption. But in several cases, the carbon was found in the carpals and digits themselves. This result was interpreted to mean that distal differentiation tendencies are acquired as a result of dedifferentiation. Since distalization was expressed only in a transplanted blastema, Faber proposed that, during regeneration *in situ,* the distal differentiation tendencies of the cone stage blastema cells are gradually converted to proximal tendencies under the influence of inductive factors emanating from the stump. The concentration of stump factors grades off distally and the hand forms autonomously from cells proliferated at the tip of the blastema under the influence of an apical organization center which arises there.

That young blastema cells somehow acquire distal morphogenetic tendencies is certainly a reality, since the progeny of such cells derived from the humeral region give rise to all structures distal to that region. But the hypothesis that young blastema cells are all initially distalized to the extent described by Faber is questionable. His conclusions revolve around the few cases in which carbon marks remained within the transplants. In the majority of his cases, however, lack of proximal development was associated with considerable resorption of blastema material. It cannot be ruled out that distalization is an artifact stemming from loss of cells which would have formed proximal skeleton had they remained within the transplant. Therefore, the possibility remains that the capacity of the young blastema for autonomous development has been underestimated.

The critical test of Faber's hypothesis would be to transplant the blastema or fractions thereof in a manner which holds dedifferentiation of the grafts to a minimum. The transplantation experiments of the present study were designed to test the extent of the self-organization capacity of the blastema as a whole, and to determine the morphogenetic behavior of blastema fractions under conditions which either maximized or minimized their dedifferentiation and resorption. The results indicate that the cells of the cone stage blastema are not distalized to the extent that they can form only hand structures in the absence of the stump. Rather, it appears that they are programmed with morphogenetic and differentiative information specifying the more proximal limb components as well and are able to express this information independently of the stump.

MATERIALS AND METHODS

All experiments were carried out on the larvae of *Ambystoma maculatum* (Shaw) raised and maintained as described previously (Stocum, 1968b). Regeneration blastemas were obtained by bilateral amputation of the forelimbs, just proximal to the elbow.

Larvae with regenerating limbs were first anesthetized in 1:1000 MS:222 (Sandoz) in spring water and then transferred to petri dishes containing MS:222 in full-strength Holtfreter solution at a concentration of 1:3000, for operation under continuous low-level anesthesia. The left forelimb was removed and the blastema carefully separated from the stump with fine iris scissors or knives (Heiss, Germany). In some cases, the regenerate of the right forelimb was left intact as an isochronous control. In other cases, it was removed as was the left regenerate and immediately fixed as an operational control. All experimental regenerates were autografted onto wound beds made in the dorsal fin by removing a small patch of skin with watchmaker's forceps. Holtfreter solution promotes wound-healing in urodeles, and healing together of the graft and fin wound borders was complete within 2–3 hours. Upon completion of wound healing, the operated animals were transferred to large finger bowls containing spring water. Transplants were allowed to develop at 21°C for 20–25 days before fixation. For each type of experiment, all three stages of regenerate (cone, palette, and notch) were transplanted. The following types of transplants, totaling 369 cases, were carried out.

Whole Blastemas Transplanted with and without Stump

Whole blastemas were transplanted with and without stump in order to determine the extent to which the blastema can undergo self-organization and to determine the effect of stump on this process. The transplants were positioned in a plane perpendicular to that of the fin, in normal proximodistal orientation. All blastemas were grafted without regard to their *in situ* dorsoventral orientation.

Blastema Fractions Transplanted without Stump

Distal half blastemas. Blastemas were halved transversely, and the distal halves were transplanted onto wound beds in a plane perpendicular to that of the fin, in normal proximodistal orientation. All grafts were made without regard to their *in situ* dorsoventral orientation.

Maximally dedifferentiated proximal half blastemas. Blastemas were halved transversely, and the proximal halves were transplanted onto wound beds in a plane perpendicular to that of the fin, in normal proximodistal orientation. The distal cut ends of such grafts became covered with fresh wound epithelium, and dedifferentiation of the transplants was maximized (see Faber, 1960, 1965). All grafts were made without regard to their *in situ* dorsoventral orientation.

Minimally dedifferentiated proximal half blastemas. A method of transplantation was desired for proximal half blastemas in which conditions favorable to migration of wound epithelium and subsequent dedifferentiation would be minimized. To this end, the following procedure was adopted. The epithelium was stripped from the ventral side of the blastema, after which the latter was halved transversely. The proximal halves were then transplanted onto wound beds with their denuded ventral surfaces facing the fin and with their distal ends oriented toward the dorsal border of the fin. The longitudinal and anterior-posterior axes of such grafts lay in the plane of the fin and the transplant remained covered externally with its dorsal epithelium. Migration of new wound epithelium was restricted to the edges of the transplant.

Minimally dedifferentiated longitudinal blastema fractions. Blastemas were cut longitudinally into anterior and posterior fractions, and the fractions were transplanted onto separate wound beds with their longitudinal cut surfaces facing the fin and with their distal ends oriented toward the dorsal border of the fin. The longitudinal axis of such a graft lay in the plane of the fin, while its anteroposterior axis lay in a plane perpendicular to the fin. Migration of new wound epithelium was again restricted to the edges of the transplant and the latter remained covered with its old epithelium.

Whole blastema transplants were either (a) fixed in Gregg's fixative and stained *in toto* for cartilage with methylene blue by the Van Wijhe method, as modified by Gregg and Butler (published in Hamburger, 1960) or (b) fixed in Bouin's solution, sectioned at 10 μ, and stained with iron hematoxylin and light green for examination of muscle. In addition, all sectioned transplants were examined for the presence or absence of lower arm skeleton. All fractional transplants were fixed in Gregg's and stained *in toto* for cartilage as described above. To gain some idea of the extent to which transplants became reinnervated by nerves from the dorsal fin, five additional whole blastema grafts were fixed in Bouin's after 5 days of development and another five after full development. Ten-micron frontal sections were prepared, stained according to the Holmes silver nitrate method, and counterstained with luxol fast blue (Margolis and Pickett, 1956).

The characteristic skeletal formula of the normal limb is 1:2:8:4:9, signifying the number of skeletal elements in the upper arm, lower arm, carpal, metacarpal, and phalangeal regions, respectively. Individual graft cartilages were easily identifiable by their distinctive morphologies and positions in relation to other cartilages. The degree of development of isochronous control and transplanted regenerates was assessed by counting the number of skeletal elements in each anatomical region of the final regenerate, and dividing by the number of elements found in the corresponding region of the normal limb.

The values obtained are percentages of the numbers of elements in the normal limb regions and will be called "indices of development." For example, if a transplanted regenerate formed six elements in the carpal region, the index of development for that region was 6/8, or 0.75 (75%).

Statistical significance of difference between mean indices of development was determined by t test for small sample size. Significance of difference was tested between mean indices of the different regions of a given graft type, between mean indices of corresponding regions of different graft types, and between mean indices of corresponding regions of isochronous control and experimental regenerates. All data were programmed professionally, and t values were computed by the appropriate method on an IBM 360/40 computer at the University of Pennsylvania Computer Center. The assistance of Mr. Peter Kuner and Computer Associates in preparing the computer program is gratefully acknowledged. The criterion for statistical significance was a $p \leqslant 0.01$. Means, standard deviations, and values for p have been recorded in a doctoral dissertation (Stocum, 1968a). The appropriate t tests and computer programs are on file at the University of Pennsylvania Computer Center.

RESULTS

Operated animals recovered from the anesthesia within half an hour after transfer to spring water and displayed normal behavior up to the time of sacrifice. Autografted regenerates became revascularized within 2–5 days after operation. Transplants of all stages developed somewhat more slowly than their isochronous controls and usually did not attain the size of the latter. Variable degrees of resorption were observed to occur in most transplants. The rate of resorption varied in individual cases, but usually ceased within 2–5 days, after which transplants increased rapidly in size. Sections of operational controls indicated that the surgical procedures employed in isolating the blastemas were effective in completely excluding stump tissues.

Reinnervation of whole blastema transplants by nerves from the dorsal fin was extremely sparse by 5 days after transplantation, although histogenesis was often well underway within the grafts. None of the five transplants examined for innervation at this time possessed more than 5–12 nerve fibers. Figure 1 illustrates two nerve fibers of a stumpless whole cone stage transplant at 5 days after operation.

The few nerve fibers present in this transplant were found only on one side of the graft and penetrated only a short distance into its basal portion. Any nerves present in a transplant at this time were always found in the form of single fibers. By 25 days after transplantation, however, nerve fibers were found in great numbers throughout individual grafts and could be traced to their terminations in muscle bundles and the epidermis. The fibers were now found both singly and in the form of fascicles.

Of the 369 total transplants, 335 (91%) developed. The remaining 34 cases underwent total resorption. Fifteen of these 34 cases were restricted to the maximally dedifferentiated proximal half category. The other 19 cases were about equally distributed among the other five transplant categories. All quantitative data are based on the 335 cases that underwent development.

The degree of development exhibited by an autografted regenerate depended to a great extent on the conditions imposed upon it by the method of transplantation. The details of development for isochronous controls and for each category of transplant are given separately in the following descriptions.

Isochronous Controls (40 Cases)

Four main skeletal regions were regenerated distal to the amputation plane: the lower arm (consisting of radius and ulna), carpals, metacarpals, and phalanges (the latter three regions collectively called the hand). Most control regenerates developed regional numbers of skeletal elements conforming to the normal formula 2:8:4:9, but in some cases the hand regions failed to form the normal numbers. Therefore, the mean indices of development of the latter regions did not quite equal 100%, but the differences from the latter value were almost negligible. Figure 2 represents an isochronous control regenerate which exhibited regional numbers of skeletal elements conforming to the normal formula.

Whole Blastemas Transplanted with Stump (79 Cases; 28 Cone, 28 Palette, and 23 Notch Stage Transplants)

As illustrated by Fig. 3, all mean regional indices of development of these transplants were essentially no different from those of isochronous controls, no matter what the stage of the regenerate at the time of transplantation. One-hundred percent of the cases examined

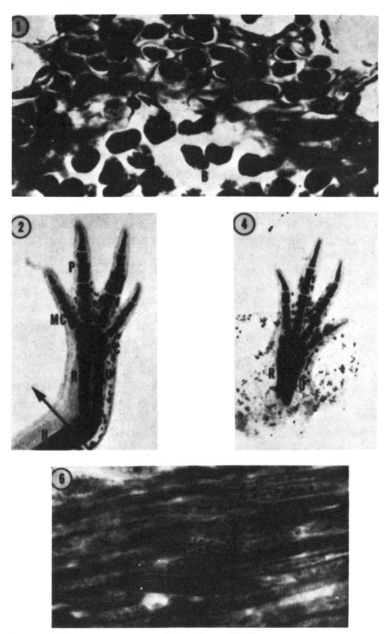

Key to abbreviations: C, carpal; *D*, digit; *H*, humerus; *MC*, metacarpal; *P*, phalange; *R*, radius; *U*, ulna; *Z*, lower arm.

FIG. 1. Nerve fibers (arrows) in epidermis (*E*) of an isolated whole cone

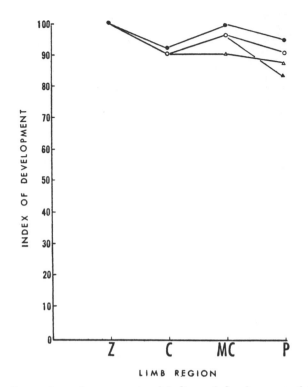

FIG. 3. Comparison of mean regional indices of development of isochronous control regenerates and whole blastemas transplanted with stump. Filled circles, control; open circles, cone stage transplant; open triangles, palette stage transplant; filled triangle, notch stage transplant.

for striated muscle had differentiated this tissue in normal anatomical relationships with the skeleton. However, the quantity of muscle differentiated was always considerably less than that of isochronous controls, and was especially sparse in the hand regions.

---

stage blastema graft 5 days after operation. The section is through a region toward the periphery, and at the base of the transplant. B, blastema cells. × 288.

FIG. 2. Twenty-five-day isochronous control regenerate of normal skeletal formula. Arrow indicates level of amputation. × 18.

FIG. 4. Twenty-five-day regenerate formed from an isolated whole cone stage transplant. Note the entirely normal morphogenesis and differentiation (skeletal formula, 2:8:4:9). × 18.

FIG. 6. Longitudinal section illustrating striated muscle differentiated in an isolated whole cone stage blastema transplant. × 288.

Whole Blastemas Transplanted without Stump (113 Cases; 39 Cone, 36 Palette, and 38 Notch Stage Transplants)

The major finding from these grafts was that numerous cases of each transplant stage were able autonomously to form the skeletal elements of all regions distal to the amputation plane. The morphology of these elements was completely normal, although fusion of cartilages was a common occurrence. Figure 4 illustrates a cone stage transplant that has formed a regenerate of normal skeletal formula.

However, the mean numbers of skeletal elements formed per transplant region never approached the values for corresponding isochronous control regions, and varied with the stage. Figure 5 represents a comparison between corresponding mean regional indices of isoch-

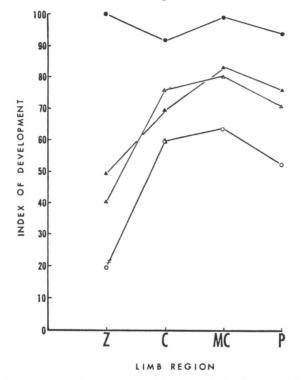

FIG. 5. Comparison of mean regional indices of development of isochronous control regenerates and whole blastemas transplanted without stump. Filled circles, control; open circles, cone stage transplant; open triangles, palette stage transplant; filled triangles, notch stage transplant.

ronous controls and transplants of each stage. No matter what the stage, the mean index of every transplant region was significantly less than that of the corresponding control region. The mean index of each transplant region increased from the cone to the notch stage, but the increase was not significant. However, for every transplant stage, the mean index of each hand region was significantly greater than that of the lower arm, although there were no significant differences among mean indices of the hand regions themselves.

Striated muscle (Fig. 6) was differentiated in normal anatomical relationships to the skeleton in 90% of cone stage transplants and in 100% of palette and notch stage transplants. There appeared to be no stage-dependent difference in the quantity of muscle formed by individual transplants, but the amounts differentiated were always far less than that in isochronous controls and slightly less than that in whole blastemas transplanted with stump.

Blastema Fractions Transplanted without Stump

Distal half blastemas (50 cases; 15 cone, 15 palette, and 20 notch stage transplants). Only skeletal elements of the hand were ever formed by distal half transplants. Figure 7 represents a comparison of mean regional indices of transplants with those of isochronous controls. For all stages, mean indices of transplants were significantly less than those of controls. The frequency of development of metacarpal and phalangeal elements in distal half grafts was comparable to that of whole blastemas transplanted without stump, but that of the carpal region was considerably less. Figure 8 illustrates the degree of development exhibited by a typical distal half graft.

Maximally dedifferentiated proximal half blastemas (18 cases; 6 cone, 7 palette, and 5 notch stage transplants). Unlike distal half transplants, the distal ends of these proximal half grafts became covered with fresh wound epithelium. Subsequently, the grafts began to regress, and eventually assumed a conical configuration by the time regression ceased. Faber (1960) has shown that the cells of partially differentiated proximal half blastema transplants of this type undergo dedifferentiation.

One of the most salient features of this transplant category was the number of grafts that underwent total resorption. Forty-five percent (5/11 total grafts done) of cone transplants, 30% (3/10 total grafts done) of palette transplants, and 58.3% (7/12 total grafts done) of

102

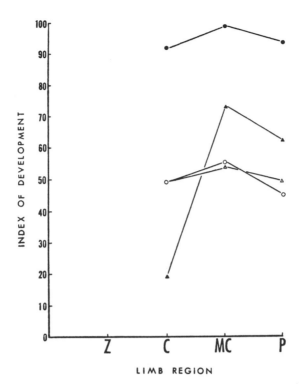

INDEX OF DEVELOPMENT

LIMB REGION

FIG. 7. Comparison of mean regional indices of development of isochronous control regenerates and distal half transplants. Filled circles, control; open circles, cone stage transplant; open triangles, palette stage transplant; closed triangles, notch stage transplant.

notch transplants were totally resorbed. All other surviving cases formed only skeletal elements of the hand, with the exception that one case each was found of cone and palette stage transplants which developed radius and ulna in addition.

Figure 9 represents a comparison of mean regional indices of transplants and isochronous controls. Indices of development were uniformly low in the transplants, and were all significantly less than those of corresponding regions of controls or whole blastemas transplanted without stump. For all stages, mean indices of the hand regions were somewhat higher, though not significantly so, than the mean index of the lower arm. Figure 10 illustrates a typical proximal half transplant which formed only digits.

Minimally dedifferentiated proximal half blastemas (38 cases; 15

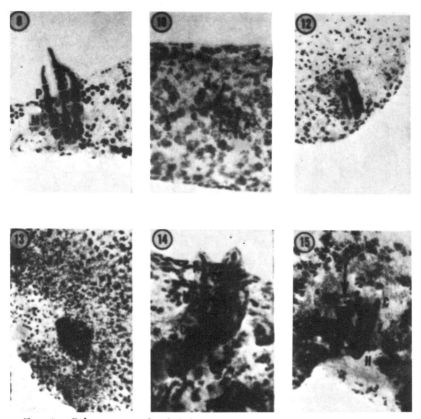

Fig. 8. Palette stage distal half transplant, illustrating exclusive tormation

Fig. 8. Palette stage distal half transplant, illustrating exclusive formation of skeletal elements of the hand.

Fig. 10. Cone stage maximally dedifferentiated proximal half transplant. Note the exclusive formation of digital skeleton.

The following six photographs represent 25-day whole mounts of fractional blastema transplants stained for cartilage with methylene blue. All figures, × 17.

Fig. 12. Exclusive development of lower arm and carpal skeleton in a palette stage minimally dedifferentiated proximal half graft. Note that the cartilages are in the plane of the fin.

Fig. 13. Exclusive formation of lower arm skeleton in a notch stage minimally dedifferentiated proximal half transplant. The cartilages are in the plane of the fin.

Fig. 14. Cone stage minimally dedifferentiated proximal half transplant which has formed both lower arm and hand skeleton. The lower arm cartilages are in the plane of the fin, while all other elements are external to the fin.

Fig. 15. Palette stage minimally dedifferentiated proximal half transplant. A short segment of humerus, both lower arm elements, and a few carpals have developed in the plane of the fin. A short digital spike (arrow) has arisen at the edge of the graft, external to the fin.

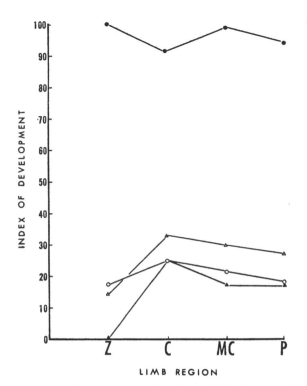

FIG. 9. Comparison of mean regional indices of development of isochronous control regenerates and maximally dedifferentiated proximal half transplants. Filled circles, control; open circles, cone stage transplant; open triangles, palette stage transplant; filled triangles, notch stage transplant.

cone, 12 palette, and 11 notch stage transplants). In contrast to maximally dedifferentiated transplants, minimally dedifferentiated proximal half grafts exhibited a high frequency of development of the lower arm skeleton. Figure 11 represents a comparison of corresponding mean regional indices of transplants and controls. For all transplant stages, the mean index of the lower arm region did not differ significantly from that of the controls, but was significantly greater than that of whole blastemas transplanted without stump.

The majority of cases formed hand skeleton in addition to that of the lower arm. For any given transplant stage, mean indices of the hand regions decreased in a proximodistal direction, and the mean index of any given hand region decreased from the cone to the notch

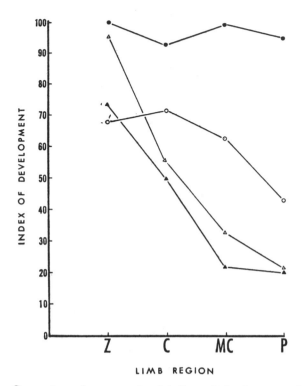

FIG. 11. Comparison of mean regional indices of development of isochronous control regenerates and minimally dedifferentiated proximal half transplants. Filled circles, control; open circles, cone stage transplant; open triangles, palette stage transplant; filled triangles, notch stage transplant.

stage. Mean indices of all transplant hand regions were significantly less than those of corresponding control regions, but did not differ significantly from corresponding mean indices of whole blastemas transplanted without stump. The mean indices of the metacarpal and phalangeal regions were less than the mean index of the lower arm for all transplant stages. The differences were not significant for the cone stage, but were significant for palette and notch stages. Thus, there appeared to be a significant preference for the formation of lower arm skeleton over hand skeleton in palette and notch transplants of this type.

A finding of major interest was that several cases were found of both palette and notch stage transplants which formed only the radial and ulnar cartilages, or these elements plus a few carpals. Figure 12 illustrates a palette stage transplant from which only the radius, ulna, and three carpals developed. Figure 13 represents a notch stage graft which formed only radius and ulna. The lower arm skeleton and carpals developed in such grafts always lay in the plane of the fin, in the orientation in which they were grafted. However, in cases where more distal skeletal elements were formed, they were found external to the fin at an angle to the radius and ulna, and the position at which they arose was always at the edge of the transplant, where the wound borders of graft and fin had healed together. Figure 14 illustrates one such cone stage transplant from which radius, ulna, carpals, and two digits developed. The palette stage transplant represented in Fig. 15 formed five carpals, radius and ulna, and a short segment of humerus in the plane of the fin. A single digital spike arose at the edge of the graft.

In summary, restriction of migration of fresh wound epithelium over proximal half transplants appears to favor the morphogenesis of proximal skeleton over distal, a result which is in contrast to what is observed in proximal half grafts were migration of wound epithelium is maximized.

Minimally dedifferentiated longitudinal blastema fractions (37 cases; 14 cone, 14 palette, and 9 notch stage transplants). Transplants of both anterior and posterior blastema fractions formed the whole spectrum of missing skeletal elements distal to the plane of amputation in numbers proportional to the size of the fraction. If the blastema was divided equally along its longitudinal axis, each half tended to form half the skeletal elements of the lower arm, carpal, metacarpal, and phalangeal regions. Figure 16 illustrates a cone stage transplant in which each half formed one digit, 2–3 carpals, and one lower arm element. Since the skeletal elements are obscured by pigment, they are reproduced in the accompanying drawing. Figure 17 illustrates a notch stage transplant in which each half formed a single digit, one or two carpals and a lower arm element.

In the majority of cases the blastema was divided unequally. This resulted in a corresponding inequality of numbers of elements produced by the fractions. Figure 18 represents a palette stage transplant in which the larger fraction formed seven phalanges, three metacarpals, the full complement of carpals and both lower arm elements. The smaller fraction produced two phalanges, two carpals

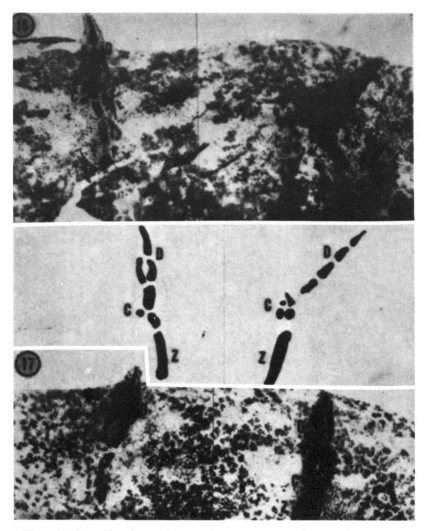

Figs. 16–19. The four photographs represent 25-day whole mounts of fractional blastema transplants stained for cartilage with methylene blue. Figs. 16 and 17, × 18; Figs. 18 and 19, × 16.5.

Fig. 16. Cone stage minimally dedifferentiated longitudinal blastema fractions of equal size. Each fraction has formed half the total number of skeletal elements produced. The elements are somewhat obscured by pigment and are reproduced in the drawing.

Fig. 17. Notch stage minimally dedifferentiated longitudinal blastema fractions of equal size. Each fraction has formed about half the total number of skeletal elements produced.

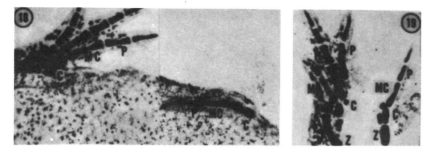

FIG. 18. Palette stage minimally dedifferentiated longitudinal blastema fractions of unequal size. Each fraction has formed numbers of skeletal elements proportionate to its size.

FIG. 19. Notch stage minimally dedifferentiated longitudinal blastema fractions of unequal size. Numbers of skeletal elements formed are in proportion to the size of the fraction.

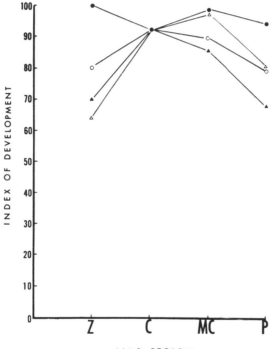

FIG. 20. Comparison of mean regional indices of development of isochronous control regenerates and minimally dedifferentiated longitudinal blastema fractions. Filled circles, control; open circles, cone stage transplant; open triangles, palette stage transplant; filled triangles, notch stage transplant.

and a single metacarpal. Figure 19 represents a notch stage transplant in which the larger fraction formed seven phalanges, 5–6 carpals, three metacarpals, and one lower arm element. The smaller fraction produced two phalanges, one metacarpal, three carpals, and a segment of the other lower arm element.

No longitudinal fraction ever tended to regulate toward the formation of a whole regenerate, although in some cases the total number of skeletal elements of a given region formed by both fractions exceeded the number characteristic of the corresponding region of the normal limb. This is considered to be due to the accidental division of the specific anlage by the surgical procedure at the time of transplantation. Figure 20 illustrates a comparison of the mean regional indices of isochronous controls with those obtained by combining the numbers of skeletal elements per region in each transplant fraction. There was no significant differences, either between corresponding regions of control and experimental regenerates, or between different regions of any given stage of experimental regenerates.

DISCUSSION

The results of the experiments described herein strongly support the concept that, by the time it has reached the cone stage, the regeneration blastema is capable of self-organization, and suggest that this capacity has been greatly underestimated by previous investigators. Two general conclusions may be drawn from the transplant data. First, transplants of isolated whole blastemas are capable at any stage of autonomously organizing into all the skeletal and muscular elements that would normally regenerate distal to the amputation plane *in situ*. Similar results have also been obtained by Jordan (1960) with conical limb blastemas of larval *Xenopus* grafted to the brain ventricle. Secondly, the various proximal and distal anatomical regions of the missing limb are represented within the blastema by the cone stage as discretely separate portions of a definitive pattern and can independently develop as such when separated in space. Distal half grafts developed into hand structures exclusively. Proximal halves of all stages of regeneration exhibited a distinct preference for the formation of proximal structures when transplanted under conditions that limited their dedifferentiation and resorption. Moreover, each transplanted longitudinal fraction of the blastema tended to autonomously organize the whole proximal-distal sequence of skeletal elements in numbers proportionate to the size of the fraction, a result which again suggests that the young blastema is a patterned structure.

The Influence of Dedifferentiation and Resorption on Development of Transplanted Regenerates

It has been noted that, although stumpless whole blastema transplants were capable at any stage of self-organizing into all the missing limb components distal to the amputation plane, they still exhibited a significant preference for the formation of hand skeleton. Faber (1960, 1965) was unable to obtain any lower arm cartilages in grafts of cone stage blastemas and interpreted this result to mean that cells of the latter possess differentiation tendencies markedly more distal than indicated by their normal fate. Another interpretation, however, would be that the apparent distalization of such grafts was the result of resorption of proximal blastema material which would have formed lower arm skeleton had it remained within the transplant. Applying this interpretation to the present results, those whole blastema transplants which underwent little or no resorption would form all skeletal elements distal to the amputation plane, while those in which large amounts of material were resorbed would form mainly hand skeleton.

However, a second factor in Faber's hypothesis was the observation that cells of a palette stage proximal half transplant, which were already differentiating to form upper and lower arm cartilages, dedifferentiated and subsequently produced mainly hand skeleton. This result has been confirmed in the present investigation with proximal half grafts designed to undergo maximum dedifferentiation. On the other hand, minimally dedifferentiated proximal half transplants exhibited a significant preference for the formation of lower arm skeleton, and the incidence of development of these elements was not significantly different from that in isochronous controls. Moreover, several of these transplants formed only lower arm (or lower arm and carpal) skeleton in the plane of the fin, and any hand cartilages that were formed arose at the edge of the grafts and at an angle to the fin. The only experimental difference between the two types of proximal half transplant was that a maximally dedifferentiated graft became covered by migrating wound epithelium, as opposed to little migration of new wound epithelium over a minimally dedifferentiated graft.

There is a great deal of evidence to indicate that the wound epithelium is involved in the dedifferentiation and accumulation of cells for the blastema (Needham, 1952; Steen and Thornton, 1963; Thornton, 1965). Moreover, the checking of dedifferentiation and the formation of a blastema on an amputated limb depends on the reestablishment of innervation to the area of the blastema (Butler and Schotté, 1941). The blastema fails to grow and resorbs if denervation of the limb is performed at any time prior to the cone stage. The blastema starts to become independent of innervation during the latter stage, since cone regenerates can sometimes proceed with more or less limited morphogenesis in the absence of nerves. Later stages of regeneration are fully independent of innervation for their continued development (Schotté and Butler, 1944; Butler and Schotté, 1949; Singer and Craven, 1948). In the present experiments, all grafts were temporarily denervated, and only a few nerves were just beginning to penetrate them by 5 days after transplantation.

Collectively, these observations suggest that the differential morphogenetic behavior of the two types of proximal half transplant described herein is due to a differential extent of epithelial-induced dedifferentiation of graft cells during the post-transplantation period when the grafts are disconnected from a nerve supply. Maximum dedifferentiation (involving the whole graft) during this period would lead to a high degree of resorption of proximal graft cells, much as a denervated amputated limb resorbs. In effect, a new plane of "amputation" would be established distal to the original amputation plane. The almost exclusive morphogenesis of hand skeleton from a maximally dedifferentiated proximal half transplant would reflect how far distally resorption had progressed.

Conversely, minimal dedifferentiation during the period of nervelessness would involve only a limited region of the graft, similar to the situation in the limb stump during normal regeneration. Those graft cells which underwent dedifferentiation would develop into structures distal to the level to which dedifferentiation had progressed, while the remainder of the graft would continue on its original course, producing more proximal structures. The number of distal skeletal elements regenerated would depend on the extent of dedifferentiation of the transplant. If no dedifferentiation occurred, only those proximal elements represented in the graft would be formed. It might be expected that progressively older graft stages would be more resistant to dedifferentiation than the cone stage and thus form progressively fewer elements of the hand skeleton. Figure 11 indicates that this is actually the case.

Distalization can thus be interpreted as an artifact associated with epithelial-induced dedifferentiation and/or resorption of graft cells in the temporary absence of innervation. Evidence supporting this interpretation has been provided by DeBoth (1965). He demonstrated that if loss of blastema material by resorption was compensated for by transplanting two or more cone blastemas to the same site, or by supplying a single transplant with a deviated nerve, the grafts were able to undergo much more complete development.

In view of the present results, it is apparent that all missing components of an amputated limb can be self-organized by the blastema, at least during the cone stage and thereafter. However, it is still possible that the stump transmits inductive messages to the blastema prior to the cone stage. In this regard, it is worth looking again at the behavior of stumpless proximal half blastemas transplanted in normal proximodistal orientation. Faber (1960, 1965) observed that the partially differentiated cells of a palette stage proximal half dedifferentiated upon transplantation. The same result was obtained when proximal halves were transplanted with reversed proximodistal orientation (Michael and Faber, 1961). In both cases, the dedifferentiated cells subsequently redifferentiated into some of the missing limb structures. The point is, that all the processes of regeneration (dedifferentiation, growth and redifferentiation, and morphogenesis) took place in the complete absence of differentiated stump tissue. Thus, a strong argument can be generated in support of the idea that the mesodermally derived stump tissues play no inductive role in the development of the blastema over its entire history. Although the asymmetry of the future regenerate could possibly be determined by the stump, it is equally possible that development of asymmetry is also an autonomous function of information programmed within the cells of the blastema from their inception. Thus, the role of the stump tissues may only be to ensure a supply of free blastema cells whose developmental information is independently expressed whether in the presence of stump or not.

SUMMARY

Whole limb blastemas of larval *Ambystoma maculatum* (Shaw) were autografted to the dorsal fin, with or without stump, in order to determine the extent of their ability to self-organize. In addition, proximal and distal fractions and longitudinal fractions of blastemas were autografted to the dorsal fin in order to determine whether the separate fractions would develop autonomously as specific parts of the total regenerate which would normally have been formed *in situ*.

Autografts of all stages of whole blastemas were able to self-organize into all the skeletal and muscular components of the lost limb parts distal to the amputation plane in the absence of the stump. Fractional parts of the blastema were able to develop autonomously into separate parts of the total regenerate corresponding to the level of the blastema from which they were obtained, providing their developmental stability was unaltered by the influence of fresh wound epithelium migrating over the graft.

It has been concluded that the blastema, at least from the cone stage on, contains a pattern of discrete and separate parts which make up a wholly self-organizing system.

The author wishes to express his appreciation to Dr. Charles E. Wilde, Jr., for his encouragement and advice through the course of this investigation, and for critical reading of the manuscript.

REFERENCES

BUTLER, E. G., and SCHOTTÉ, O. E. (1941). Histological alterations in denervated non-regenerating limbs of urodele larvae. *J. Exptl. Zool.* 88, 307–341.

BUTLER, E. G., and SCHOTTÉ, O. E. (1949). Effects of delayed denervation on regenerative capacity in limbs of urodele larvae. *J. Exptl. Zool.* 112, 361–392.

DAVID, L. (1932). Das Verhalten von Extremitätenregeneraten des weissen und pigmentieren Axolotl bei heteroplastischer, heterotopen und orthotopen Transplantation und sukzessiver Regeneration. *Roux' Arch. Entwicklungsmech. Organ.* 126, 457–511.

DEBOTH, N. J. (1965). Enhancement of the self-differentiation capacity of the early limb blastema by various experimental procedures. *In* "Regeneration in Animals" (V. Kiortsis and H. A. L. Trampusch, eds.), pp. 420–426. North Holland Publ., Amsterdam.

FABER, J. (1960). An experimental analysis of regional organization in the regenerating forelimb of the axolotl (*Ambystoma mexicanum*). *Arch. Biol.* 71, 1–67.

FABER, J. (1965). Autonomous morphogenetic activities of the amphibian regeneration blastema. *In* "Regeneration in Animals" (V. Kiortsis and H. A. L. Trampusch, eds.), pp. 404–418. North Holland Publ., Amsterdam.

HAMBURGER, V. (1960). "A Manual of Experimental Embryology," revised ed., p. 196. Univ. of Chicago Press, Chicago, Illinois.

JORDAN, M. (1960). Development of regeneration blastemas implanted into the brain. *Folia Biol.* (*Kraćow*) **8**, 41–53.

MARGOLIS, G., and PICKETT, J. P. (1956). New applications of the luxol fast blue myelin stain. *Lab. Invest.* **5**, 459–474.

METTETAL, C. (1939). La régénération des membres chez la salamandre et le Triton. Histologie et détermination. *Arch. Anat. Histol. Embryol.* **28**, 1–214.

METTETAL, C. (1952). Action du support sur la differenciation des segments proximaux dans les régénérats de membre chez les Amphibiens Urodèles. *Compt. Rend. Acad. Sci.* **234**, 675.

MICHAEL, M. I., and FABER, J. (1961). The self-differentiation of the paddle-shaped limb regenerate, transplanted with normal and reversed proximal-distal orientation after removal of the digital plate (*Ambystoma mexicanum*). *Arch. Biol.* (*Liege*) **72**, 301–330.

NEEDHAM, A. (1952). "Regeneration and Wound-Healing," pp. 62–64. Methuen, London.

NEEDHAM, J. (1942). "Biochemistry and Morphogenesis," pp. 430–442. Cambridge Univ. Press, London and New York.

PIETSCH, P. (1961). Differentiation in regeneration. I. The development of muscle and cartilage following deplantation of regenerating limb blastemata of *Ambystoma* larvae. *Develop. Biol.* **3**, 255–264.

SCHOTTÉ, O. E., and BUTLER, E. G. (1944). Phases in regeneration and their dependence on the nervous system. *J. Exptl. Zool.* **97**, 95–122.

SINGER, M., and CRAVEN, L. (1948). The growth and morphogenesis of the regenerating forelimb of adult *Triturus* following denervation at various stages of development. *J. Exptl. Zool.* **108**, 279–308.

STEEN, T. P., and THORNTON, C. S. (1963). Tissue interaction in amputated aneurogenic limbs of *Ambystoma* larvae. *J. Exptl. Zool.* **154**, 207–222.

STOCUM, D. L. (1968a). Doctoral Dissertation, Univ. of Pennsylvania, Philadelphia, Pennsylvania.

STOCUM, D. L. (1968b). The urodele limb regeneration blastema: A self-organizing system. I. Differentiation *in vitro*. *Develop. Biol.* **18**, 441–456.

THORNTON, C. S. (1965). Influence of the wound skin on blastema cell aggregation. *In* "Regeneration in Animals" (V. Kiortsis and H. A. L. Trampusch, eds.), pp. 333–339. North Holland Publ., Amsterdam.

DEVELOPMENT OF THE
NEUROMUSCULAR JUNCTION

I. Cytological and Cytochemical Studies on the Neuromuscular Junction of Differentiating Muscle in the Regenerating Limb of the Newt *Triturus*

THOMAS L. LENTZ

INTRODUCTION

Following amputation of the limb of the newt, muscle and other specialized tissues of the limb dedifferentiate to form a blastema composed of mesenchymal cells (Butler, 1933; Thornton, 1938; Singer, 1952; Chalkley, 1954; Hay and Fischman, 1961; Hay, 1962; Trampusch and Harrebomée, 1965; Lentz, 1969). Subsequently, the undifferentiated mesenchymal cells differentiate to reform muscle, bone, and connective tissue. Thus, during limb regeneration, muscle undergoes a complete developmental cycle in which all of its differentiative states are represented. This system affords an opportunity to correlate the degree of specialization of the neuromuscular junction with the developmental stage of the muscle cell.

This paper describes reinnervation of differentiating muscle and formation of the neuromuscular junction in the newt. In addition, the appearance and localization of acetylcholinesterase (AChE) activity was determined histochemically to correlate enzymatic specialization with morphological development. Other studies have been performed on development of the neuromuscular junction in the rat (Kelly, 1966; Terävänien, 1968) and chick (Hirano, 1967). As a prior step to this investigation, the cytological aspects of muscle dedifferentiation and differentiation in the newt limb have been described in detail (Lentz, 1969). It will be shown here that both morphological and chemical (AChE) specializations characteristic of the neuromuscular junction develop only where axon endings and muscle become closely apposed. The present findings have been reported briefly (Lentz, 1968).

MATERIALS AND METHODS

Adult newts, *Triturus viridescens*, used in these experiments, were maintained in the laboratory in large aquaria. For study of motor end-plate development, newts were anesthetized in 1% chloretone and the limb was amputated at the level of the lower third of the upper forelimb. The newts were then placed on damp paper in covered finger bowls, fed chopped beef liver three times a week, and the limb was allowed to regenerate for periods of 6–12 wk. Although muscle differentiation begins as early as 2 wk after limb transection, the later time periods were selected because all stages in muscle differentiation from mesenchymal cell to fully differentiated muscle can be found in the regenerate at this time. After about 8 wk, more advanced stages of differentiation predominate. The newts were then anesthetized, and the regenerating limb was amputated for fixation. Muscle and overlying epidermis of the regenerate was trimmed from the cartilage and cut into small blocks. The tissue blocks were placed in cold 3% glutaraldehyde in 0.05 M cacodylate buffer (pH 7.2) and fixed for 1 hr. The blocks were rinsed briefly

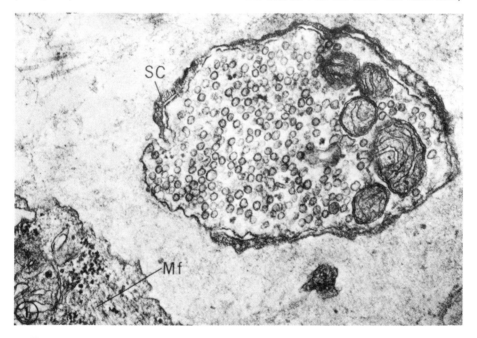

FIGURE 1 A vesicle-filled axon is located in the intercellular space near a differentiating muscle cell containing myofilaments (*Mf*). The Schwann cell (*SC*) covering is deficient over the side of the axon facing the muscle cell. No changes have occurred on the muscle surface opposite the axon. 6-wk regenerate. × 58,500.

in cold buffer and fixed for an additional hour in cold 1% osmium tetroxide in 0.05 M cacodylate buffer at pH 7.2. The tissues were dehydrated rapidly in ethanol, infiltrated with Maraglas (Freeman and Spurlock, 1962), and polymerized at 49°C. Thick, 1–2 μ sections were cut with glass knives on a Porter-Blum MT-1 microtome and stained with 0.1% toluidine blue for light microscopic orientation. Thin sections were cut from blocks containing muscle, stained with lead hydroxide (Feldman, 1962), and examined with an RCA EMU 3F electron microscope.

The thiolacetic acid–lead nitrate method (Crevier and Bélanger, 1955; Barrnett and Palade, 1959; Barrnett, 1962) was chosen for the demonstration of cholinesterase in the developing motor end plate. Comparison of this method with several other available procedures has shown that it results in a sharp and accurate localization at the electron microscopic level (Bloom and Barrnett, 1966; Koelle and Gromadzski, 1966; Davis and Koelle, 1967). A limitation of the method is its low specificity, which can be obviated to some degree by the use of appropriate inhibitors.

Tissue was obtained in the same manner for the histochemical investigations, rinsed in buffer after glutaraldehyde fixation, and placed in the incubating media for 60 min at 4°C. The incubating media contained 0.0012 M thiolacetic acid (Eastman Organic Chemicals, Rochester, New York; redistilled in the laboratory) and 0.006 M Pb(NO3)2 in cacodylate buffer at a final pH of 6.8. After incubation, the tissues were rinsed in three changes of buffer, 5 min each. They were then placed in osmium tetroxide and prepared for electron microscopy in the manner described above.

Control procedures consisted of omission of substrate from the medium or, alternatively, addition of physostigmine (eserine) (10^{-4} M) or diisopropylfluorophosphate (DFP) (10^{-5} M and 10^{-7} M) to the incubating medium. The tissues were placed in buffer containing inhibitor for 20 min prior to exposure to the incubating medium with inhibitor. In the case of eserine, the buffer rinse prior to osmium tetroxide fixation contained inhibitor.

RESULTS

Cytology of Neuromuscular Junction Formation

Structures comparable to axon terminations are first seen in close relationship to muscle only when differentiation of the latter has progressed to the stage of a multinucleate cell containing myofibrils. These axons are dilated and contain some small

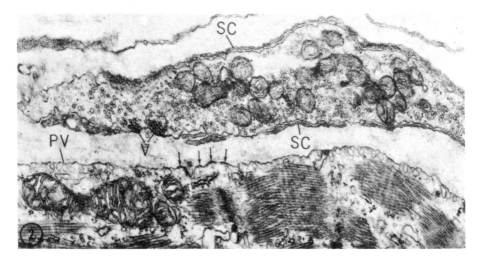

FIGURE 2   Axon containing vesicles and mitochondria near a differentiating muscle cell in a 6-wk regenerate. The Schwann cell (*SC*) covering is complete over the outer surface of the fiber, but is discontinuous over the side facing the muscle cell. A few slight elevations or ridges (arrows) with underlying densities occur on the muscle surface. Pinocytotic vesicles (*PV*) are not so abundant on the muscle surface facing the center of the nerve fiber termination as they are laterally. In the nerve fiber, there are two clusters of vesicles (*V*) beneath projections from the surface. × 26,000.

mitochondria, large numbers of small vesicles (500 A) with contents of low density, and smaller numbers of large vesicles with dense contents (Fig. 1). The Schwann cell cytoplasm is deficient or incomplete with gaps in it over the side of the axon facing the differentiating muscle cell but continues to envelop the opposite side (Figs. 1, 2).

The first morphological changes indicative of motor end-plate formation develop at the apposing surfaces of differentiating muscle and axon terminal where the Schwann cell covering is incomplete (Fig. 2). On the muscle surface, a few low elevations or ridges occur (Fig. 2). There is a slight increase in density of the cytoplasm immediately beneath the plasma membrane of the elevations. Pinocytotic vesicles are not so common along the muscle surface facing the axon as over the rest of the muscle surface. On the surface of the axon, elevations or evaginations similarly project outward, but there is no cytoplasmic density associated with them. Synaptic vesicles, abundant throughout the nerve terminal, seem more concentrated adjacent to the plasma membrane facing the muscle cell and are tightly packed beneath the elevations on the axon surface (Fig. 2).

In ensuing stages, the axon becomes more

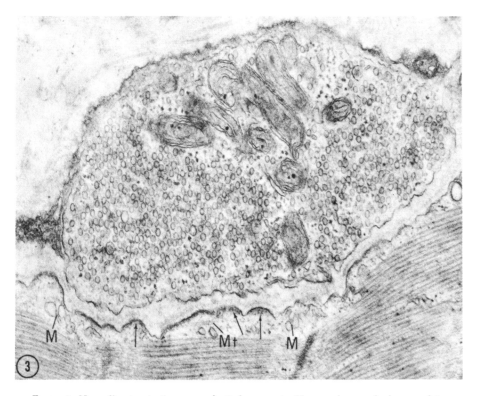

FIGURE 3 Nerve fiber termination on muscle. 6-wk regenerate. The axon is more closely apposed to the muscle cell than in the previous stage, and its contour roughly follows that of the muscle surface. Schwann cell cytoplasm is now completely absent over the inner surface, and the filamentous coating of muscle (basement lamina) occupies the intercellular space. The ridges (arrows) on the muscle surface are more prominent, and dense material is accumulated on the inner aspect of the sarcolemma of the folds. Pinocytotic vesicles are diminished in the region of the junction while some irregularly shaped membranous structures (*M*) and a microtubule (*Mt*) occupy the sarcoplasm between myofilaments and plasma membrane. × 43,500.

closely approximated to the muscle cell, coming to lie in a shallow depression or gutter on the surface of the muscle (Fig. 3). Schwann cell cytoplasm is now completely absent on the side of the axon-facing muscle. The contour of the ending generally follows that of the muscle, producing a relatively constant synaptic cleft. The cleft is occupied by filamentous material that has a denser central band somewhat closer to the muscle than to the axon surface. At the same time, the ridges on the muscle surface become more numerous and prominent and can definitely be identified as early junctional folds (Fig. 3). The subplasmalemmal densities are also more apparent and are due to the accumulation of fine-textured material (Fig. 3).

Subsequently, the junctional folds increase in length and the intercellular space is reduced further (Fig. 4). The bases of the folds are nar-

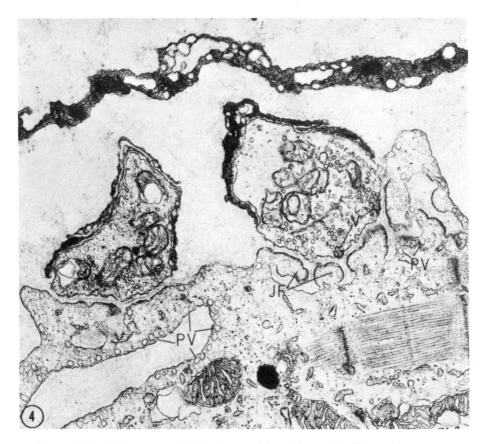

FIGURE 4   Developing neuromuscular junction on a differentiating muscle cell in a 6-wk regenerate. The axon is closely apposed to the muscle cell, with the densest part of the filamentous intercellular material being situated roughly equidistant between nerve and muscle membranes. In the axoplasm, vesicles are segregated toward the muscle. The right hand portion of the junction appears more advanced than the region to the left. In the former, junctional folds (*JF*) are apparent; in the axon, vesicles (*V*) accumulate beneath projections opposite the clefts between junctional folds. Pinocytotic vesicles (*PV*) are absent in the immediate region of the junction except at the bases of the folds, but are numerous along the rest of the muscle surface. Membranous structures and tubulovesicles are found in the sarcoplasm beneath the junction. × 37,500.

120

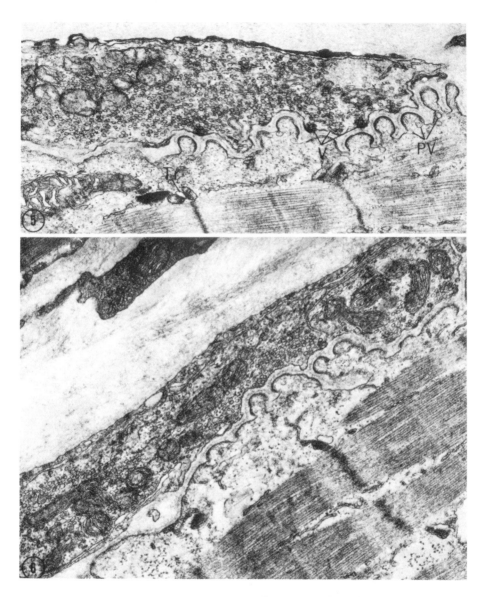

FIGURE 5 Neuromuscular junction from a limb allowed to regenerate for 6 wk. Junctional folds are well developed and have pinocytotic vesicles (PV) at their bases. The intercellular space is ~600 Å across and occupied by a band of filamentous material. Vesicles (V) are abundant in the axon and are especially concentrated beneath projections of the nerve surface toward the secondary clefts. This junction is morphologically mature and indistinguishable from normal junctions (Fig. 6). Note also that the muscle cell is highly differentiated, containing well organized myofibrils with triads (Tr) associated with them. × 28,500.

FIGURE 6 Motor end plate from the upper forelimb of a normal newt. Compare the structural features of this junction and the sarcoplasm with those in Fig. 5. × 29,000.

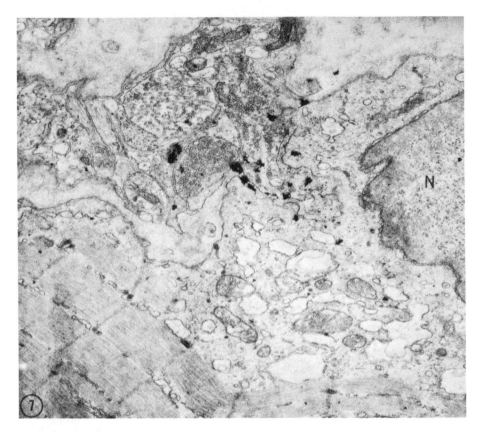

FIGURE 7 Low magnification of an early neuromuscular junction reacted for cholinesterase activity. Note that enzymatic activity is largely restricted to the region of close contact between the vesicle-filled axon termination and the muscle cell surface. Junctional folds are emerging in the same area. A few reaction deposits are present in the axoplasm of the nerve fiber and in the enveloping Schwann cell cytoplasm. Irregular membranous structures (some with associated activity), mitochondria, and small granules resembling ribosomes occur in the subjunctional sacroplasm. Myofibrils are found outside this region. N, nucleus. × 17,500.

rower than the distal region, giving the folds a club-shaped appearance. With the closer apposition of axon to muscle, the dense band in the filamentous material occupying the synaptic cleft is situated roughly equidistant between adjacent plasma membranes. The subplasmalemmal cytoplasmic density in the muscle cell is found only in the junctional folds, being absent in other regions of the junction, for example, at the bases of the clefts between folds. Pinocytotic vesicles are largely restricted to the bases of the junctional folds and are rarely seen along the sides and ends of the folds (Fig. 4). Outside the end-plate region, pinocytotic vesicles in association with the sarcolemma are numerous. In the axon terminal, synaptic vesicles are segregated in the pole adjacent to the muscle cell. Focal accumulations of synaptic vesicles immediately beneath the axon membrane are more common (Fig. 4). These tightly packed masses of vesicles occur where the axon contour projects slightly opposite the secondary synaptic clefts. In the early stages of junction formation, only a small amount of sarcoplasm separates the myofibrils from the sarcolemma (Figs. 1–3). This area increases in size as development proceeds, and it contains irregularly shaped membranous profiles, an occasional rough-surfaced cisterna, mitochondria, glycogen granules, free ribosomes, and sparsely distributed microtubules (Figs. 3, 4).

Increase in the area of axon contact with muscle results in the formation of myoneural junctions (Fig. 5) indistinguishable from those observed in normal animals (Fig. 6). The fully formed motor end plates are found on muscle cells that are also fully differentiated, or nearly so. Myofibrils, previously short and separated by wide stretches of cytoplasm, are now long and tightly packed. The membranous elements of the sarcoplasm (transverse tubules and sarcoplasmic reticulum) are also developed and oriented in the usual relation to each other and the myofibrils (e.g., triad in Fig. 5).

Appearance of Cholinesterase Activity

The final reaction product (PbS) of the thiolacetic acid–lead nitrate method for cholinesterase occurs as discrete, electron-opaque particles. The particles are variable in shape and size, depending on the amount of final product precipitated, but can be accurately localized to subcellular structures and organelles (Bloom and Barrnett, 1966; Davis and Koelle, 1967). Because the final product is precipitated at or near the site of substrate hydrolysis, the number of deposits would seem to depend on the number of enzymatic sites. Thus, it is not surprising that relatively few reaction deposits are observed in the early stages of junction formation while more are seen in the later developmental stages.

Enzyme activity was not found on intercellular nerve fiber endings or on the muscle surface prior to the appearance of morphological indications of motor end-plate formation. Enzymatic activity was demonstrated only where axon terminations are closely approximated to the muscle surface and where junctional folds are beginning to emerge (Fig. 7). Physostigmine- and DFP-sensitive activity appears in the following locations: within membranous structures in the muscle cytoplasm, on the plasma membranes of muscle and axon, and within the axoplasm of the terminations.

In the muscle cell, deposits of final product are found on small membranous tubules or vesicles (tubulovesicles) (Figs. 8, 9). Although membranous structures of this type occur all along the cortical sarcoplasm, only those subjacent to the end-plate region contain activity. Some of these tubulovesicles containing reaction product appear to be connected with or to have fused with the plasma membrane (Fig. 8). Serial sections show that some of the apparently isolated tubulovesicles beneath the plasma membrane are actually connected to the membrane (Figs. 8, 9). These tubulovesicles seem to differ from the pinocytotic invaginations that occur along the muscle surface and at the bases of the junctional folds; they are variable in size, not as regular in shape, and not restricted to the bases of the folds. Other small deposits, not enclosed by a vesicle, are applied to the inner aspect of the plasma membrane (Fig. 8).

Larger deposits of final product occur on the

FIGURES 8 and 9 Serial sections through a developing neuromuscular junction in an 8-wk regenerate reacted for cholinesterase activity. Corresponding deposits of final reaction product in the two figures are numbered. The smallest deposits occur in membranous structures or tubulovesicles beneath or applied to the inner surface of the sarcolemma (Fig. 8, *1, 2, 3, 4, 8, 10, 15, M*; Fig. 9, *1, 2, 4, 8, 10, 13, M*). Comparison of the two figures shows that some of the apparently isolated subsarcolemmal deposits are connected to the plasma membrane (*1, 2, 4, 10*) and that those on the inside of the membrane may project into the intercellular cleft (*3, 13, 15*). Some of the small vesicles containing reaction product are fused with the plasma membrane (Fig. 8, *2*; Fig. 9, *4*). Larger reaction deposits occur on the outer surface of the sarcolemma and may fill the intercellular cleft (*5, 6, 13, 14*). Activity is present on both the inner (*12*) and outer (*11*) surfaces of the nerve membrane and also in the axoplasm. A dense dot occurs in the center of some of the synaptic vesicles. A few small deposits occur in the Schwann cells enveloping the axon and in a membrane-bounded structure, resembling a lysosome, within the process of a connective tissue (*CT*) cell. × 26,000.

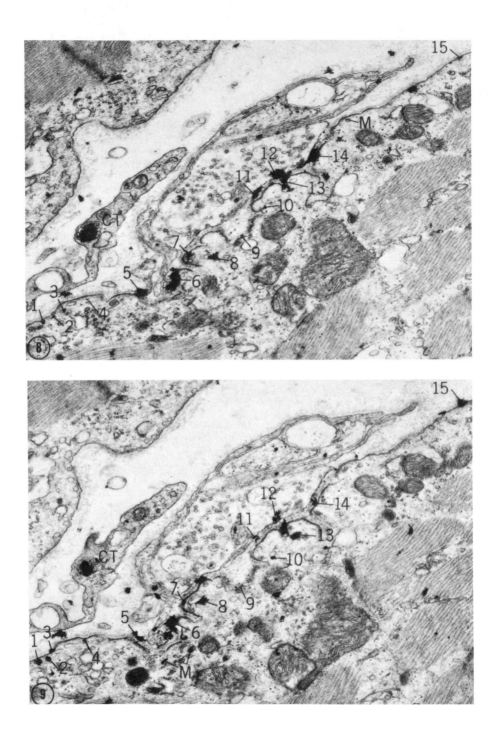

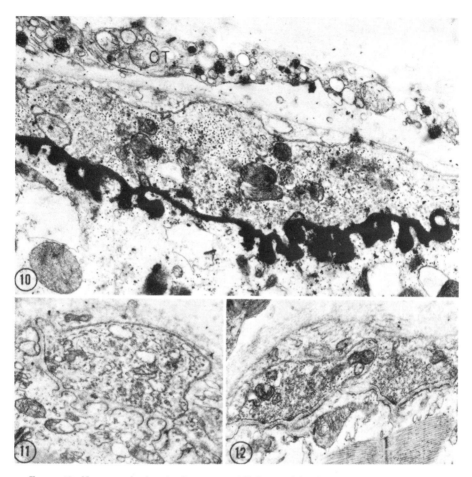

FIGURE 10 Neuromuscular junction from a normal limb reacted for cholinesterase activity. Activity in this junction is unusually intense, with reaction product filling the junctional clefts. Lysosome-like bodies in the connective tissue (*CT*) cell process overlying the nerve terminal are reactive. × 17,000.

FIGURE 11 Highly developed motor end plate preincubated in physostigmine (10^{-4} M) and incubated in the reaction media for cholinesterase which also contained physostigmine. 8-wk regenerate. The reaction in the end-plate region has been almost completely inhibited. Most of the persisting lead deposits occur in the intercellular spaces around the junction. × 16,000.

FIGURE 12 Neuromuscular junction after incubation in the reaction media for AChE which contained DFP (10^{-5} M). Activity in the motor end plate is inhibited. Lead deposits do occur in association with extracellular collagen fibrils and dispersed over the myofibrils. 8-wk regenerate. × 19,500.

outer surface of the sarcolemma (Figs. 8, 9). Most of these are located at the tips of the ridges and more fully developed junctional folds. Some deposits are so large that they extend across the intercellular cleft, but the serial sections show that most of them originate on the muscle surface (Figs. 8, 9).

Reaction product is less abundant on the axon

125

surface and is localized to both sides of the axon membrane (Figs. 8, 9). The deposits on the inner side of the membrane could not be associated with vesicles or other structures. Deposits on the outer axon surface are not numerous and not so large as those on the outer surface of the sarcolemma. Activity also occurs in the axoplasm, but could not be associated consistently with a specific organelle (Figs. 8, 9).

As development of the motor end plate progresses, the reactive sites on the plasma membrane increase in size and number. Although morphologically well developed junctions have a normal localization of enzyme activity, the activity is not so intense, as evidenced by the number of reaction deposits, as in motor end plates of normal animals (Fig. 10), even after 12 wk of regeneration. In normal animals, the reaction is sometimes so intense as to fill the intercellular space and structures such as tubulovesicles, pinocytotic vesicles, and synaptic vesicles that might be continuous with the space.

Occurrence of reaction product at the sites described above was almost completely prevented by preincubation of the tissues in physostigmine (10^{-4} M) (Fig. 11) and DFP (10^{-5} M) (Fig. 12). Inhibition of activity at these sites occurred at all stages of end-plate formation. Activity was only partially inhibited, however, by DFP at a concentration of 10^{-7} M. Small lead deposits in Schwann cell cytoplasm (Figs. 7, 8) were not significantly affected by inhibitors and are presumably due to an enzyme other than a cholinesterase. Large membrane-bounded granules, resembling lysosomes, in connective tissue cells are reactive (Figs. 8, 10). This activity was resistant to inhibition and is probably also due to another enzyme, possibly cathepsin C. A small dense deposit was variably observed in synaptic vesicles of both experimental and control incubations. Incubation in media without substrate sometimes produced small dense deposits throughout the section, especially over myofilaments and collagen fibrils. These deposits could be reduced or eliminated, however, by adequately rinsing the tissues in buffer prior to osmium fixation.

DISCUSSION

Morphogenetic Relationships

In differentiating muscle, both morphological (e.g., junctional folds) and chemical (AChE)

specializations characteristic of the neuromuscular junction develop only where axon terminals become closely apposed to the muscle surface. In no cases were postjunctional structures or localized concentrations of enzyme activity found on the muscle surface outside the area of contact. Thus, these findings support the conclusion of others (Zelená, 1962; Kelly, 1966; Teräväinen, 1968), that motor nerve fibers exert a morphogenetic effect on the formation of the subneural apparatus.

It was observed here that nerve and muscle are already differentiated to some extent before endplate formation begins. As noted by Hirano (1967), muscle differentiation progresses through the myoblast and early myofiber stages before neuromuscular junctions are formed. The intercellular nerve fiber terminations change from growing end bulbs that invade the early regenerate to expansions filled with synaptic vesicles. The early end bulbs differ from the vesicle-filled terminals of later regenerates in that they have fewer small vesicles, but more membranous channels and tubules and larger vesicles containing dense material (Lentz, 1967). Junctions develop only in relation to the vesicle-filled endings. If either the axon or muscle is less differentiated, the motor end plate does not form, even if the axon passes near the muscle cell. Formation of the neuromuscular ujnction, then, depends on a relationship between nerve fiber and muscle that have each attained a required degree of specialization. If the inductive interaction between nerve and muscle is linked to a mechanism of mutual recognition which depends on transfer of information, it seems likely that a certain degree of specialization is necessary for transfer and/or reception of the information.

A significant question involves the time of appearance of cholinesterase activity in relationship to innervation of the muscle cell. Some investigators have found that cholinesterase appears before innervation and suggest that prior appearance of the enzyme may have a chemotactic influence on the approaching motor nerve fiber (Kupfer and Koelle, 1951; Beckett and Bourne, 1958; Shen, 1958). Other investigators have concluded that concentration of enzyme activity at the end-plate region occurs only after axon contact with muscle has been established (Mumenthaler and Engel, 1961; Zelená, 1962; Khera and Laham, 1965; Kelly, 1966). The present findings support the latter conclusion and show that cholinesterase activity develops concurrently with morphological differentiation. This conclusion is supported further by studies of muscle cells in tissue culture; the uninnervated cells do not develop cholinesterase-containing motor end-plate structures (Engel, 1961). On this evidence, it appears that the motor nerve fiber directly influences the formation of the chemical, as well as the morphological, specialization of the neuromuscular junction.

Cytological Specializations

The most conspicuous event in motor end-plate development is the formation of the junctional folds. The folds differ somewhat in structure from the adjacent muscle surface, indicating that they may form by outgrowth from, rather than folding of, the muscle surface. Pinocytotic vesicles, abundant all along the muscle surface, are not associated with the membrane of the folds but persist only at their bases. Similarly, the cytoplasm of the folds is largely devoid of structures such as ribosomes or glycogen granules found throughout the rest of the cytoplasm. On the other hand, dense material, not found elsewhere, accumulates beneath the plasma membrane of the fold. Dense material has also been described in the junctional folds of intrafusal motor myoneural junctions of the frog (Karlsson and Andersson-Cedergren, 1966). The subsarcolemmal densities in the motor end plate are similar in appearance to accumulations of material beneath the postsynaptic membrane of synapses from several areas including the central nervous system (Bloom and Aghajanian, 1966). Although the exact nature of this material is unknown, the material is thought to be proteinaceous and presumably is involved in synaptic function. If this is the case also for the motor end plate, its appearance is significant as a further indication of acquisition of functional capacity.

Enzymatic (AChE) Specialization

In the earliest stages of junction formation, enzymatic activity is associated with tubulovesicles in the muscle cytoplasm only in the region beneath the axon. These reactive structures also seem to be a component of the normal motor end plate (Miledi, 1964). Some of the tubulovesicles are connected to the sarcolemma, indicating their possible role in delivery of the enzyme from deeper levels to the cell surface. A somewhat similar process has been suggested for the electroplaque of the eel in which reactive tubulovesicles containing mucoid, extracellular-space substance are continuous with the surface (Bloom and Barrnett, 1966). The reverse process, inward migration of the reactive sites by a process of pinocytotic invagination, does not seem as likely because of the sparse surface localization of enzyme at this time and the fact that unreactive pinocytotic vesicles can be distinguished from the reactive membranous structures. Larger deposits are found in association with the sarcolemma, indicating accumulation of enzyme at this site.

The origin of AChE within the muscle cell is unknown. Structures such as mitochondria, glycogen, a few rough-surfaced cisternae of endoplasmic reticulum, and free ribosomes are found in the sarcoplasm near the developing end plate. No unusual or special relationships were found between these structures and demonstrable activity that might suggest a role in enzyme formation. Ribosomes are associated with protein synthesis, and their persistence in the end-plate region (Padykula and Gauthier, 1967) while decreasing in the rest of the muscle cell during development (Lentz, 1969) might implicate them as sites of synthesis.

This work was supported by a grant (GB-7912) from the National Science Foundation and by grants (TICA-5055, National Cancer Institute, and FR-5358) from the National Institutes of Health, United States Public Health Service.

REFERENCES

BARRNETT, R. J. 1962. The fine structural localization of acetylcholinesterase at the myoneural junction. *J. Cell Biol.* **12**:247.

BARRNETT, R. J., and G. E. Palade. 1959. Enzymatic activity in the M band. *J. Biophys. Biochem. Cytol.* **6**:163.

BECKETT, E. B., and G. H. BOURNE. 1958. Some histochemical observations on enzyme reactions in goat foetal cardiac and skeletal muscle and some human foetal muscle. *Acta Anat.* **35**:224.

BLOOM, F. E., and G. K. AGHAJANIAN. 1966. Cytochemistry of synapses: selective staining for electron microscopy. *Science.* **154**:1575.

BLOOM, F. E., and R. J. BARRNETT. 1966. Fine structural localization of acetylcholinesterase in electroplaque of the electric eel. *J. Cell Biol.* **29**:475.

BUTLER, E. G. 1933. The effects of X-radiation on the regeneration of the fore limb of *Amblystoma* larvae. *J. Exp. Zool.* **65**:271.

CHALKLEY, D. T. 1954. A quantitative histological analysis of forelimb regeneration in *Triturus viridescens. J. Morphol.* **94**:21.

CREVIER, M., and L. F. BÉLANGER. 1955. Simple method for histochemical detection of esterase activity. *Science.* **122**:556.

DAVIS, R., and G. B. KOELLE. 1967. Electron microscopic localization of acetylcholinesterase and nonspecific cholinesterase at the neuromuscular junction by the gold-thiocholine and gold-thiolacetic acid methods. *J. Cell Biol.* **34**:157.

ENGEL, W. K. 1961. Cytological localization of cholinesterase in cultured skeletal muscle cells. *J. Histochem. Cytochem.* **9**:66.

FELDMAN, D. G. 1962. A method of staining thin sections with lead hydroxide for precipitate-free sections. *J. Cell Biol.* **15**:592.

FREEMAN, J. A., and B. O. SPURLOCK. 1962. A new epoxy embedment for electron microscopy. *J. Cell Biol.* **13**:437.

HAY, E. D. 1962. Cytological studies of the dedifferentiation and differentiation in regenerating amphibian limbs. *In* Regeneration. D. Rudnick, editor. Ronald Press, New York. 177.

HAY, E. D., and D. A. FISCHMAN. 1961. Origin of the blastema in regenerating limbs of the newt *Triturus viridescens. Develop. Biol.* **3**:26.

HIRANO, H. 1967. Ultrastructural study on the morphogenesis of the neuromuscular junction in the skeletal muscle of the chick. *Z. Zellforsch.* **79**:198.

KARLSSON, U., and E. ANDERSSON-CEDERGREN. 1966. Motor myoneural junctions on frog intrafusal muscle fibers. *J. Ultrastruct. Res.* **14**:191.

KELLY, A. M. 1966. The development of the motor end plate in the rat. *J. Cell Biol.* **31**:58A. (Abstr.)

KHERA, K. S., and Q. N. LAHAM. 1965. Cholinesterase and motor end-plates in developing duck skeletal muscle. *J. Histochem. Cytochem.* **13**:559.

KOELLE, G. B., and C. G. GROMADZSKI. 1966. Comparison of the gold-thiocholine and gold-thiolacetic acid methods for the histochemical localization of acetylcholinesterase and cholinesterase. *J. Histochem. Cytochem.* **14**:443.

KUPFER, C., and G. B. KOELLE. 1951. A histochemical study of cholinesterase during formation of the motor end plate of the albino rat. *J. Exp. Zool.* **116**:397.

LENTZ, T. L. 1967. Fine structure of nerves in the regenerating limb of the newt *Triturus. Amer. J. Anat.* **121**:647.

LENTZ, T. L. 1968. Cytological and cytochemical studies of development of the neuromuscular junction during limb regeneration of the newt *Triturus. J. Cell Biol.* **39**:154A. (Abstr.)

LENTZ, T. L. 1969. Cytological studies of muscle dedifferentiation and differentiation during limb regeneration of the newt *Triturus. Amer. J. Anat.* **124**:447.

MILEDI, R. 1964. Electron-microscopical localization of products from histochemical reactions used to detect cholinesterase in muscle. *Nature.* **204**:293.

MUMENTHALER, M., and W. K. ENGEL. 1961. Cytological localization of cholinesterase in developing chick embryo skeletal muscle. *Acta Anat.* **47**:274.

PADYKULA, H. A., and G. F. GAUTHIER. 1967. The ultrastructure of neuromuscular junctions of mammalian red and white skeletal muscle fibers. *J. Cell Biol.* **35**:155A. (Abstr.)

SHEN, S. C. 1958. Changes in enzymatic pattern during development. *In* The Chemical Basis of Development. W. D. McElroy and B. Glass, editors. Johns Hopkins Press, Baltimore. 416.

SINGER, M. 1952. The influence of the nerve in regeneration of the amphibian extremity. *Quart. Rev. Biol.* **27**:169.

TERÄVÄINEN, H. 1968. Development of the myoneural junction in the rat. *Z. Zellforsch.* **87**:249.

THORNTON, C. S. 1938. The histogenesis of muscle in the regenerating forelimb of larval *Amblystoma punctatum. J. Morphol.* **62**:17.

TRAMPUSCH, H. A. L., and A. E. HARREBOMÉE. 1965. Dedifferentiation a prerequisite of regeneration. *In* Regeneration in Animals and Related Problems, V. Kiortsis, and H. A. L. Trampusch, editors. North-Holland Publishing Co., Amsterdam. 341.

ZELENÁ, J. 1962. The effect of denervation on muscle development. *In* The Denervated Muscle. E. Gutmann, editor. Publishing House of the Czechoslovak Academy of Science, Prague. 103.

The Fine Structure of Tissues in the Amputated-Regenerating Limb of the Adult Newt, *Diemictylus viridescens* [1]

WESLEY P. NORMAN [2]   AND ANTHONY J. SCHMIDT

The amputated forelimb of the adult newt, *Diemictylus viridescens* regenerates and, in doing so, exhibits the biological processes of inflammation, cell and tissue modulation, blastema development, growth, and differentiation (see Schmidt, '66 for review). The composite of these biological activities is a new limb that replaces the lost appendage.

Regeneration as a whole is the result of cooperation among several heterogeneous processes on a temporal basis, and phases may be distinguished that lend themselves to individual investigation. A preblastemic phase includes wound-closure and healing of the amputee. This is followed successively by a blastema phase, and then a differentiative and morphogenetic phase leading to a reconstituted limb. Of these phases, the formation of the blastema, beginning in the preblastemic phase from among the injured limb tissues, and the

growth of the blastema itself are absolute prerequisites for regeneration. Therefore, much of our attention has been devoted to the fine structure of injured tissues and cells of the preblastemic, and blastemic phases; the differentiating tissues were selectively investigated.

MATERIALS AND METHODS

The adult newts, *Diemictylus viridescens*, obtained from the vicinity of Petersham, Massachusetts, were prepared for this investigation by first dehumerizing their forelimbs (see Schmidt, '62 for details), and amputating above the elbow three weeks later. The experimental ani-

[1] This investigation was supported by a research grant GM 06208, and a graduate training grant 8T1-HD-16, from the Public Health Service.
[2] This paper was prepared in part from a thesis submitted in partial fulfillment of the requirements for the Degree of Doctor of Philosophy.

mals were placed into their own finger bowls containing aerated distilled water, and maintained in incubators at 20°C for the duration of the study.

Postsurgical treatment with Aqua Aid (a fungicide, Longlife Fish Food Products, Harrison, N. J.) prevented, or markedly reduced infection of the wounds. Excepting the first week following surgery, the animals were fed Tubifex every two weeks, and their containers refreshed with clean water biweekly.

The animals were anesthetized with MS-222 (Tricaine Methanesulfonate, Sandoz Pharmaceuticals, N. Y.) for all surgical procedures, including the final sampling of the regenerating limbs. Regenerates were sampled at representative phases (see Schmidt, '66) during an interim of 35 days following amputation. The limbs were fixed in 5% glutaraldehyde buffered at pH 7.9–8.2 with 0.1 M Veronal (Sabatini et al., '63) for 16 hours, at 4°C. The fixed limbs were rinsed with cold Veronal buffer, bisected, and placed in buffered (0.1 M Veronal, pH 7.6) ice-cold osmium tetroxide (Palade, '52) for one hour. The limbs were subsequently prepared for embedding in Epon (Luft, '61).

Sections of about 500–900 Å were cut on a Porter-Blum MT-2 ultramicrotome, picked up on 150 and 200 mesh copper grids coated with parlodion and carbon, and stained with uranyl acetate (Huxley and Zubay, '61; Watson, '58), and lead citrate in sodium hydroxide (Reynolds, '63). Observations were made on the Hitachi HU 11A electron microscope.

More extensive details of our procedure may be found in a previous paper (Norman and Schmidt, '67) as well as in the studies cited above.

RESULTS

The sequence of regeneration of the adult newt limb has been briefly summarized in our introduction. These phases will serve to subdivide our presentation of the fine structure of the amputated regenerating limb tissues, with emphasis placed on the wound epithelium, peripheral nerves, striated muscle, and cells forming the blastema.

Preblastemic phase

During the first three days following amputation, the cells and tissues of the limb undergo both degenerative and reparative structural alterations. The first reparative change is that of the migrating epidermal cells over the amputation surface. The wound surface is covered by epithelial cells as early as 4–6 hours following amputation, and, by 24 hours, the epithelium has attained a thickness of eight to ten cells.

The basal epithelial cells at the margin of the wound do not show marked fine-structural changes. Cytoplasmic structures and nuclei are not appreciably different than those observed in normal epidermis (see Norman and Schmidt, '67). Typically found are round and elongated mitochondria, granular endoplasmic reticulum, clusters of free ribosomes, and bundles of filaments. The basal cell inner surface may show marked irregularities (fig. 1), and many vesicles within the cytoplasm. The glycocalyx[4] is continuous beneath the basal cells and follows the contour of the irregular basal cell inner surface. Hemidesmosomes are fewer than in the normal epidermis. The plasma membranes of adjacent basal cells are highly folded. These cytoplasmic folds, or projections, are moderately electron-dense, and usually contain no cellular organelles. The basal epithelial cells over the center of the wound appear flattened during the early phases of regeneration, and possess a more regular border than the cells at the wound margin; hemidesmosomes are entirely lacking (fig. 2).

The fine structure of cells forming the wound epithelium appears not to differ significantly from that of the cells at the edge of the wound. However, some exceptions are noted, particularly in cells that appear to be phagocytic in function (fig. 3). The cell illustrated in figure 3 contains many nuclear and cytoplasmic fragments, as well as the usual cytoplasmic organelles. Desmosomal junctions, typical of the wound epithelium, are quite obvious between this cell and neighboring cells. In

---

[4] The term "glycocalyx" was introduced by Bennett ('63) to describe the polysaccharide-protein coating or "basement membrane" found around cells and adjacent to epithelia.

addition, the wound epithelium has been observed to contain many pyknotic nuclei and a considerable amount of cellular debris within polymorphonuclear leucocytes (fig. 14), and macrophages of histogenetic origin. These cells wander through the intercellular spaces of the epithelium. Desmosomal contacts between these wandering cells and the epithelial cells have not been observed.

Among the subepithelial tissues, the most obvious change during the first phase of regeneration is the loss or dissolution of the adepidermal reticulum of fibers. The organization of the reticulum is lost (fig. 1); collagen fibrils fragment, and disappear altogether (fig. 2). With the loss of this adepidermal reticulum, subjacent cells are able to make close contact with the basal cells of the wound epithelium. Figure 2 illustrates a macrophage in very close proximity to the cytoplasmic projections of an epithelial cell (large arrow).

During the period immediately following amputation and lasting through the early preblastemic phase of regeneration, the most numerous cells under the wound epithelium are those of vascular origin. The cellular components are red blood cells, monocytes, lymphocytes, and polymorphonuclear leucocytes. With the exception of the polymorphonuclear leucocytes, these cells disappear by the third through fifth day after amputation. Macrophages migrate through the wound fluid where they apparently phagocytose red blood cells (fig. 2) and other cellular debris.

The nerve bundles early reveal an obvious response to the trauma of limb amputation. These nerve bundles consist of both myelinated and nonmyelinated axis cylinders, Schwann cells, and connective tissue sheaths. Degenerative changes are evident in the myelinated nerves by 24 hours following injury. Within a single bundle of nerve fibers, large numbers of axons and their investing myelin sheaths undergo marked changes. The injured axons are highly vacuolated and, in some cases, have lost their integrity altogether (fig. 4). Myelin sheaths are disorganized and osmiophilia is lost in irregular patches (see also Schmidt, '66a). Still, scattered among these degenerating injured nerves are normal-appearing nonmyelinated axons (fig.

4). The degenerating nerve bundles lack a well organized perineurium (fig. 5) and, in some cases, the perineurium is completely lost. Furthermore, the components of a nerve bundle tend to separate from one another during the early days of regeneration. This separation or looseness is particularly obvious in regions of the nerve bundles just proximal to the plane of injury (fig. 6). Individual Schwann cells appear to be isolated from one another and to enfold only a few axons. Some of the axons are enclosed by myelin. Large pigment-containing cells have also been observed within nerve bundles, associated with degenerating nerve fibers (fig. 4).

The major tissue of the deboned forelimb of the newt consists of striated muscle. Although regeneration of striated muscle has been frequently investigated in adult urodeles, there have been few reports on the fine structure of alterations that take place during the early stages of repair. It is important to recognize that all muscle fibers have not been equally injured by the transection of a muscle bundle during the amputation of the limb. Following amputation, the characteristic striated appearance of the terminal portions of the injured muscle fibers disappears. This loss of myofiber striations is probably related to two simultaneous processes. Some myofibers perish and, in necrosis, obviously lose their characteristic myofibrillar structure through the dissolution of the myofilaments and the Z-bands. The first structures to be lost are the thick (myosin) filaments. The myofiber illustrated in figure 7 is typical of a cell which has undergone irreparable damage: this is a muscle fiber in which the innermost fibrillar structures are lost, and possess a pyknotic nucleus. The only structures still relatively intact are the mitochondria; however, even these organelles are swollen and have lost their typical electron-dense matrix. Mitochondria are often gathered into large clusters in the terminals of these injured muscle fibers.

When death of a muscle fiber occurs, the cytoplasmic and nuclear debris is phagocytized by the many macrophages migrating into the surrounding connective tissue sheaths of the muscle and under the glycocalyx of degenerating myofibers. A portion of a macrophage with ingested muscle debris is illustrated in figure 8. In many cases, the only remaining identifiable structure of what was once a muscle fiber is a highly tortuous glycocalyx (fig. 9). The glycocalyx apparently resists destruction and remains intact amidst a fibrous environment: these glycocalyxes are found near the distal injured terminals of muscle fibers.

Other muscle fibers located in the area of amputation apparently do not undergo an irreversible change following injury. Many, if not a majority, of the muscle fibers contract away from the wound region, some leaving distal fragments behind. Thin nucleated fibers have been observed among the mysial sheaths in the distal regions of the injured muscle (fig. 10). The nuclei of these thin fibers contain finely dispersed chromatin and one or two well defined nucleoli. The nuclear membrane is made up of an inner and outer membrane. The inner membrane is interrupted by nuclear pores. The outer nuclear membrane is only partially coated with ribosomal particles (fig. 10, large arrows). This thin muscle fiber shows a prominent reaction to injury within its sarcoplasm; this reaction is a disorganization of the remaining myofilaments. Some thick (myosin) and thin (actin) filaments are lost; however, all of the thin filaments that remain intact retain densities that may correspond to the Z-bands. Triads also remain intact (fig. 13), but the remainder of the sarcoplasmic reticulum becomes disorganized and vacuolated. Mitochondria show no particular structural changes in these cells. The intact plasma membrane is studded with inpocketings, and the external glycocalyx, also intact, is highly tortuous.

Normal-appearing muscle fibers are frequently observed alongside degenerating fibers (fig. 11). In figure 11, a degenerating fiber displays a vacuolated sarcoplasmic reticulum, isolated clumps of myofilaments, swollen mitochondria and scattered clumps of glycogen particles. Distal portions of some injured muscle fibers lack bundles of myofilaments, contain mainly small mitochondria, clusters of free ribosomes, and an occasional lysosome. It is not known whether this signifies an early stage of degeneration or of repair. The bulk of the fiber is made up of a reticular network of fine filaments (fig. 12). Lamellar profiles resembling those of a Golgi apparatus have been observed in the cytoplasm of other injured but not completely degenerate cells (fig. 13).

Closely associated with the degenerating fiber of figure 11 are cytoplasmic processes of two cells incorporated within the glycocalyx of the muscle fiber. Although these cells lie within the glycocalyx, they are separated from the muscle fiber by an intervening space. These cells are in a location that resembles that of the muscle "satellite" cells described by Mauro ('61) and Church, Noronha and Allbrook ('66). The cytoplasm of these "satellite" cells contains numerous free ribosomes, and mitochondria with few cristae but dense matrices. A few flattened cisternae of endoplasmic reticulum are observable in one of the cells in figure 11. The intercellular space between myofibers of the amputated newt limb is made up, predominantly, of finely dispersed granular electron-dense material and a few scattered collagen fibrils.

Blastemic phase

From the tenth to the fourteenth day of regeneration, fine-structural changes of the wound epithelium, and the underlying nerves and muscle do not appreciably differ from those observed earlier. The blood clot has dissolved and most of the cellular and extracellular debris has been removed, particularly, by macrophages. Retrogressive nerve and muscle degeneration continues into the stump of the limb. Also, normal-appearing axis cylinders have been observed among the subapical tissues at this time. Cells begin to accumulate under the apical wound epithelium to form the primary regeneration blastema.

Cells continue to migrate distad in the amputee, adding to the *in loco* population of mitotically active cells under the wound epithelium. The growing regeneration blastema is bounded at its outer border by the wound epithelium, and at its inner border by a prominent fibrocellular reticulum. During the ensuing nine days of regeneration (15–24 days following amputation), the blastema grows from a bulb into a

132

conic protuberance: histogenesis is apparent in the cone blastema.

During this period, the wound epithelium is usually 8–10 cell layers thick, and closely resembles the normal epidermis in fine structure. The basal cells present a regular free border, and a glycocalyx is evident (fig. 14). A homogeneous granular material adjacent to the glycocalyx is continuous with the intercellular matrix of the blastema proper. With the growth of the blastema into a conic protuberance by about the twenty-second day after amputation, palisading of the basal cells of the wound epithelium becomes increasingly apparent. Hemidesmosomes appear along the free surface of these basal cells (fig. 15), the subjacent glycocalyx increases in electron density and, underneath, lies a fibrillar feltwork that antecedes the regenerating adepidermal reticulum of the normal dermoepidermal junction. These fibrils are randomly disposed within a granular ground substance, and also extend haphazardly into the intercellular space of the adjacent blastema (see also Salpeter and Singer, '59, '60).

The cells forming the regeneration blastema appear quite similar to each other when observed with the light microscope. Indeed, with few exceptions, a similarity among these cells is also evident when viewed with the higher resolution of the electron microscope. The blastema is made up of cells possessing many of the characteristics of fibroblasts (figs. 16, 17). The nuclei of these cells are large, and their chromatin is condensed, principally along the inner surface of the nuclear membrane. There are a few cells in which the nuclear chromatin is obviously more condensed than is evident among the majority of blastema cells. One or two nucleoli are usually present within the nucleus, and the nuclear membrane is studded with pores. The fine structure of the cytoplasm of these cells is fairly uniform from cell to cell. Small mitochondria, oftentimes swollen, are scattered throughout the cytoplasm. The internal structure of the mitochondria consists of only a few cristae projecting into a moderately dense matrix. In most cells, an extensive Golgi apparatus is clearly evident, to one side of the nucleus (fig. 16). Profiles of rough-surfaced endoplasmic reticulum are found generally near or around the nucleus, and are less frequently seen in the extensions of these cells. The cytoplasmic extensions of the blastema cell contain numerous free ribosomes (fig. 17). Lysosomes, and lipid droplets, are also observed in these cells. The plasma membrane of the blastema cells is typically irregular, and even indistinct in some regions; glycocalyx-like condensations of ground substance lie adjacent to the outer surface of the plasma membrane. Between the cells is a granular intercellular material containing a few nondescript fibers, and a scattering of banded collagen fibrils. Rarely, plasma membranes of neighboring blastema cells are thickened (desmosome-like), forming adhesion plates; such thickenings are observed between cells in a young blastema of a 17-day regenerate (fig. 20).

In the late blastema (22–24 days), some cells become flattened, an early sign of organization into a cartilage anlagen of the forelimb skeleton. When observed, these cells (prochondroblasts) are surrounded by a moderate amount of dense granular matrix (fig. 18). However, when closely applied to each other, very little matrix intervenes between these cells (fig. 18, small arrows): these cells are the youngest prochondroblasts distinguishable from surrounding blastema cells. The cytoplasm of these prochondroblasts retains many of the characteristics of blastema cells of origin. Profiles of rough-surfaced endoplasmic reticulum are prominent and clusters of free ribosomes are evident (figs. 18, 19) in the cytoplasm. The free ribosomes are found in quantity, only in the outer margins of the cell cytoplasm or in cytoplasmic extensions. An adhesion plate similar to the one found in younger (17-day) regenerates has also been observed in a 22-day regenerate (fig. 21).

Deep within the blastema and forming a limiting boundary between it and the underlying stump tissue, is a structurally distinct fibrocellular region. This fibrocellular region has been particularly obvious during the fifteenth to twenty-second day of regeneration around the terminals of muscle and nerve tissues. In this region, the muscle terminals contain typical striated fibrils (figs. 22, 23). These muscle

133

cells display numerous mitochondria with typical cristae and a dense matrix. Scattered among the mitochondria are dense bodies resembling lysosomes. Numerous glycogen particles are scattered throughout the cell, as are bundles of thin filaments attached to Z-bands. Well formed triads are associated with some of these Z-bands. To date, we have found no evidence of a distinct sarcoplasmic reticulum in these cells.

The plasma membrane of regenerating muscle is normal in appearance, studded with inpocketings, and completely surrounded by a glycocalyx. In some muscle fibers, the distal cell border is highly irregular, with numerous projections (fig. 22): the plasma membrane along some of these projections shows thickenings. The plasma membrane of these projections contains a few scattered invaginations, but only in areas that are not thickened. The extracellular space about these muscle terminals contains many fibrils.

Within the region of the muscle terminals just described, lie many normal-appearing fibroblasts, and a few lipid-engorged cells (fig. 23). The fibroblastic cells lie within a region of active fibrillogenesis, and differ from similar cells in other connective tissues in that many contain collagen-like fibrils within their cytoplasm. This observation is unique not only with respect to connective tissue of vertebrates in general, but has not been found elsewhere within the regenerating adult newt forelimb. Figure 24 illustrates one such fibroblast that contains both filaments (located near the plasma membrane) and banded fibrils, within the cytoplasm. In some regions where the plasma membrane usually appears indistinct (arrow, fig. 24), one can get the impression that the intracellular fibrils are continuous with those in the extracellular space. The periodicity of the intracellular fibrils averages 570 Å. The average width of these fibrils is about 400 Å. The lengths vary, but in the cell processes of some fibroblasts, collagen fibrils as long as 4 μ have been measured (fig. 25). These measurements are consistent with those determined for the fibers in the surrounding extracellular space. Intracellular fibrils are typically found in groups of 2 or 3, partly enclosed by smooth

membranes and partly free in the cytoplasm. The collagen fibrils within the cell of figure 25 appear to be lying freely in the cytoplasm. A few are observed to be partly bounded by membranes at their distal ends (arrow, fig. 25). Filaments can be observed in one of the smaller extensions. Characteristically, there is an abundance of perinuclear rough-surfaced endoplasmic reticulum in these cells. A paucity of rough-surfaced endoplasmic reticulum found in the extensions of these cells is also characteristic of the cytoplasmic extensions of fibroblasts. A cross-section through a cell extension reveals the lack of rough-surfaced endoplasmic reticulum, the presence of numerous polysomes, and groups of fibrils cut both transversely and obliquely (fig. 26). The plasma membrane appears indistinct in some regions and, here again, one gets the impression that the fibrils are being deposited in the extracellular space.

In additional observations on these fibroblasts, banded fibrils have been found adjacent to the nucleus as well as close to the plasma membrane. Figure 27 shows numerous groups of banded fibrils lying close to the nucleus as well as scattered through the cytoplasm and near the cell membrane. Smooth membranes can be distinguished around a few of the fibril groups.

Groups of fibrils that have been cut transversely are illustrated in figure 28: the diameters of these intracellular fibrils measure about 400 Å. Figure 28 also directs attention to the proximity of some of the fibrils to Golgi vesicles and of others to the plasma membrane (arrows).

Differentiative and morphogenetic phase

Cells of the cone blastema differentiate into both chondroblasts and perichondral cells (fig. 29). The prochondroblasts observed at an earlier stage (fig. 18) are now incorporated into a definitive skeleton. A gradation of maturation is observed from the cells forming the perichondrium to the cells located deep within the cartilage. The perichondral cells are flattened, elongate, and possess fine-structural characteristics that do not differ markedly from the chondroblasts. The perichondral cells are surrounded by a fibrous matrix. In contrast, the chondroblasts are surrounded by both a reticulate collagenous and polysaccharide matrix. The chondroblasts are identified by their large nucleus and relatively—small amount of cytoplasm. The cytoplasm characteristically contains small mitochondria and a large amount of rough-surfaced endoplasmic reticulum. The cell borders are highly irregular and are surrounded by a lighter staining reticular matrix. In more active areas of cartilage formation, the irregular granularity of the matrix is more obvious. Figure 29 illustrates a chondroblast located in a region intermediate between the perichondrium and centrally located chondrocytes.

Other histogenesis within the regenerating adult newt limb requires greater attention than it has been allocated to date, and will be the subject of subsequent reports.

DISCUSSION

Precedence for some of our observations with the electron microscope may be found in studies by Salpeter and Singer ('59, '60a,b, '62). Accordingly, we will dwell only briefly on a few complementary observations, while most of our attention will focus on data not previously reported.

Membrane specializations at points of adhesion, or tight junctions between cells are frequently recognized as desmosomes. Desmosomes have been observed in the urodele, between cells of the epidermis, myoepithelial cells, and neural-sheath cells (Norman and Schmidt, '67). These structures have also been found between cells forming the wound epithelium, and between cells in the regeneration blastema. The latter contacts appear similar to those reported by Salpeter ('65) between cells and nerve terminals in the blastema; we cannot reach the same conclusion with our limited observations to date. Such cell contacts have also been observed in fetal rat tendon, fetal bovine ligamentum nuchae, and rat embryo tooth buds (Ross and Greenlee, '66), all mesenchymal tissues. Moscona ('57) has described cell contacts in embryonic tissue of like functional potentialities: this concept has been discussed in detail by Weiss ('58). Accordingly, since tight contacts have been observed between blastema cells, which are essentially cells in a stage of development, these cells may be considered to have common functional potentialities.

Hemidesmosomes are found along the inner free surface of the basal epidermal cells (see Norman and Schmidt, '67), but are absent from the basal cells of the preblastemic wound epithelium. This has been observed particularly in the wound epithelium overlying the apex of the stump, in a region where the glycocalyx and adepidermal reticulum are also obviously absent. Since hemidesmosomes are sites of epidermal attachment to the underlying tissue, their absence may reflect the degree of cell modulation necessary for epithelial migration over the injured stump tissues. The loss of such surface specializations may also allow more surface area exposure for movement of materials across the cell membrane. During the early preblastemic phase of regeneration, the basal epithelial cells are in direct contact with the wound fluid formed from tissue hydrolyses as well as plasma exudation: materials from this metabolite-rich fluid could be used for the maintenance and development of a functional wound epithelium. We have also observed signs of phagocytic activity by the wound epithelial cells (see also Singer and Salpeter, '61).

The fine structure of the blastema cells in the regenerating larval urodele limb has been described by Hay ('58) and, in the adult urodele, by Salpeter and Singer ('60, '62). Observations in the present investigation have confirmed those by Salpeter and Singer ('60, '62), and warrant only brief additional discussion here. The cells forming the regeneration blastema in the adult newt showed some structural diversity. The majority of these cells possessed a well developed granular endoplasmic reticulum, whereas others had only isolated fragments, that were swollen or irregularly shaped. Free ribosomes were usually found in cell processes. Still other cells were observed to have no endoplasmic reticulum at all. These variations are interpreted as representative of functional differences among these cells. These differences may be due to the reaction of these cells to a new environment and changes in their metabolic machinery preparatory to progressive regenerative activities, i.e., histodifferentiation into the cartilaginous skeleton of the reconstituting limb. We can conclude that the cells forming the regeneration blastema show structural characteristics that differ from cells of the limb stump tissues.

During the course of these studies, intracellular collagen-like banded fibrils were repeatedly observed in fibroblasts located deep within the regeneration blastema of 17-day regenerates (see Norman and Schmidt, '66). Electron microscopic evidence of intracellular fibrillogenesis is extremely limited, although from time to time this possibility has been suggested from observations with the light microscope (e.g., Jordan, '39; Lewis, '17; and others). These observations do not agree with the generally accepted process of collagen formation by tropocollagen polymerization in the extracellular space (Chapman, '61; Porter, '64; Revel and Hay, '63; and others). Meek ('66) has reported on collagen-like fibrils within capsule cells around neuronal ganglia of a snail, while Welsh ('66) has found fibrils within fibroblasts populating a human desmoid fibroma. In other studies, Sheldon and Kimball ('62) observed banded collagen-like fibers within vacuoles in chondroblasts, while Usuku and Gross ('65) have reported on the presence of collagen fibrils within intracellular vacuoles of cells in metamorphosing tadpoles. These latter investigators suggest that these intracellular fibrils are present as a result of phagocytosis, but do not deny the possibility of collagen synthesis contributing to the presence of these fibrils within cells. Our observations lead us to propose that the synthesis of collagen accounts for the intracellular fibrils within the regenerating adult newt limb and, very likely, elsewhere.

The cells under consideration here have been identified as fibroblasts by the presence of cytoplasmic organelles characteristic of fibroblasts associated with connective tissues. Extensive profiles of granular endoplasmic reticulum, a prominent Golgi apparatus, and peripheral cytoplasmic filaments are basic criteria for the identification of fibroblasts (Ross, '64). For the most part, the cells are in proximity to large quantities of extracellular collagen fibers. Cells other than fibroblasts that are frequently found in connective tissues contain cytoplasmic features which readily distinguish them from fibroblasts, e.g., macrophages differ from fibroblasts in their relative lack of granular endoplasmic reticulum, and the presence of particularly numerous lysosome bodies. These criteria all assist in discriminating the fibroblast from other cell types that may be found in connective tissues.

Although the active mechanism of fibril production cannot be ascertained by observing static micrographs, certain proposals are possible. In normal wound healing or in chondrogenesis, the rate of collagen production is rapid, and synthetic activities have been described and generally accepted by a number of investigators (Carneiro and Leblond, '59; Fernando and Movat, '53; Revel and Hay, '63; Ross, '64): tropocollagen is produced in the cisternae of granular endoplasmic reticulum, either packaged in vesicles of Golgi origin for externalization, or directly transported to the cell surface for expulsion into the extracellular space for final aggregation into fibrils at the cell membrane or some distance away. In a region of extremely rapid fibrillogenesis, such as apparently takes place in the depths of the regeneration blastema, is it not conceivable that the synthesis of tropocollagen is similar to that in other organisms, although here a premature aggregation of these protein filaments can occur within the cell? The preformed collagen-like fibrils may then approach the cell membrane for externalization.

However, the present evidence for intracellular fibril aggregation allows us to consider at least three modes of ontogeny. One possibility concerns the origin of the membranes surrounding the intracellular collagen fibrils. These smooth membranes may actually be invaginations of the plasma membrane, producing an extracellular environment conducive to rapid aggregation of tropocollagen units as soon as they reach the cell surface. However, the membranes around the collagen-like fibrils do not have the trilaminar structure typical of plasma membranes; there is no ectoplasm with tropocollagen filaments about these membranes, as is typical subjacent to the plasma membrane of active fibroblasts; and, the uniformity of the fibrils themselves does not compare with the band periodicity nor with the irregular diameters of the extracellular fibrils.

A second possibility suggests that the origin of the smooth membranes around some of the intracellular fibrils may be derived from the Golgi membranes. The length of a Golgi vesicle would have to increase considerably in order to enclose the long fibrils observed in some cells. Still, such lengths may be reached by tensions placed on flattened Golgi vesicles by the elongating fibrils. Therefore, the characteristically extensive Golgi apparatus found in fibroblasts may, just possibly, be able to enclose intracellular fibrils. However, there may be a simpler explanation for the intracellular fibril structure than involving a packaging and transport mechanism via the Golgi apparatus.

The third possibility, and the one that we propose, is simple, direct, and begins with the synthesis of tropocollagen within the granular endoplasmic reticulum. Following this protein synthesis, the ribosomes may leave the membranes of the endoplasmic reticulum. The resulting smooth membranes either deteriorate, or fuse with the plasma membrane to exteriorize their contents for extracellular aggregation into banded collagen fibrils. However, should there be an intracellular fusion between the tropocollagen-filled vesicles and polysaccharide-containing Golgi vesicles (the Golgi apparatus is recognized as a site of polysaccharide synthesis (Fewer, Threadgold and Sheldon, '64; Godman and Lane, '64; Neutra and Leblond, '66a,b; and others)), the protein filaments may condense into fibrils before passing out into the extracellular space. This concept warrants, and is receiving, additional study.

Finally, we should like to briefly speculate on the importance of our observations on collagen synthesis to regeneration in the adult newt. The collagen-rich environment in the regenerating limb has been localized to the region of the blastema adjacent to the stump musculature. On the basis of recent studies by Konigsberg and Hauschka ('65), Hauschka and Konigsberg ('66), who found that the differentiation of cultured myoblasts was a response to collagenous "conditioned medium," we postulate that the fibrillogenesis deep within the regeneration blastema serves to condition the extracellular space for the regeneration of striated muscle (and possibly other histogenesis as well). This thesis is receiving continuing attention in our laboratory.

LITERATURE CITED

Bennett, H. S. 1963 Morphological aspects of intracellular polysaccharides. J. Histochem. Cytochem., 11: 14–23.

Carneiro, J., and C. P. Leblond 1959 Role of osteoblasts and odontoblasts in secreting the collagen of bone and dentin, as shown by radioautography in mice given tritium-labelled glycine. Exp. Cell Res., 18: 291–300.

Chapman, J. A. 1961 Morphological and chemical studies of collagen formation. I. The fine structure of guinea pig granulomata. J. Biophys. Biochem. Cytol., 9: 639–651.

Church, J. C. T., R. F. X. Noronha and D. B. Allbrook 1966 Satellite cells and skeletal muscle regeneration. Brit. J. Surg., 53: 638–642.

Fernando, N. V. P., and H. A. Movat 1963 Fibrillogenesis in the regenerating tendon. Lab. Invest., 12: 214–229.

Fewer, D., J. Threadgold and H. Sheldon 1964 Studies on cartilage. V. Electron microscopic observations on the autoradiographic localization of S^{35} in cells and matrix. J. Ultrastructure Res., 11: 166–172.

Godman, G. C., and N. Lane 1964 On the site of sulfation in the chondrocyte. J. Cell Biol., 21: 353–366.

Hauschka, S. D., and I. R. Konigsberg 1966 The influence of collagen on the development of muscle clones. Proc. Nat. Acad. Sci., 55: 119–126.

Hay, E. D. 1958 The fine structure of blastema cells and differentiating cartilage cells in regenerating limbs of Amblystoma larvae. J. Biophys. Biochem. Cytol., 4: 583–592.

Huxley, H. E., and G. Zubay 1961 Preferential staining of nucleic acid-containing structures for electron microscopy. J. Biophys. Biochem. Cytol., 11: 273–296.

Konigsberg, I. R., and S. D. Hauschka 1965 Cell and tissue interactions in the reproduction of cell type. In: Reproduction: Molecular, Subcellular, and Cellular. M. Locke, ed. Academic Press, N. Y., pp. 243–290.

Luft, J. H. 1961 Improvements in epoxy resin embedding methods. J. Biophys. Biochem. Cytol., 9: 409–414.

Mauro, A. 1961 Satellite cell of skeletal cell fibres. J. Biophys. Biochem. Cytol., 9: 493–494.

Meek, G. A. 1966 Intracellular collagen fibres. J. Physiol., 182: 3P–4P.

Moscona, A. 1957 Formation of lentoids by dissociated retinal cells of the chick embryo. Sci., 125: 598–599.

Neutra, M., and C. P. Leblond 1966a Synthesis of carbohydrates of mucus in the Golgi complex as shown by electron microscope radioautography of goblet cells from rats injected with glucose-H³. J. Cell Biol., 30: 119–136.

——— 1966b Radioautographic comparison of the uptake of galactose-H³ and glucose-H³ in the Golgi region of various cells secreting glycoproteins or mucopolysaccharides. J. Cell Biol., 30: 137–150.

Norman, W., and A. J. Schmidt 1966 The intracellular localization of banded collagen fibrils in fibroblasts of the regenerating forelimb of the adult newt, Diemictylus viridescens. Anat. Rec., 154: 395.

——— 1967 The fine structure of limb tissues of the adult newt, Diemictylus viridescens. J. Morph., 123: 251–270.

Palade, G. E. 1952 A study of fixation of tissues for electron microscopy. J. Exp. Med., 95: 285–297.

Porter, K. R. 1964 Cell fine structure and biosynthesis of intercellular macromolecules. Biophys. J., 4: 167–196.

Revel, J. P., and E. D. Hay 1963 An autoradiographic and electron microscope study of collagen synthesis in differentiating cartilage. Zeitschr. f. Zellforsch., 61: 110–144.

Reynolds, E. S. 1963 The use of lead citrate at high pH as an electron-opaque stain in electron microscopy. J. Cell Biol., 17: 208–212.

Ross, R. 1964 Studies of collagen formation in healing wounds. In: Advances in the Biology of Skin, vol. 5. Wound Healing. W. Montagna and R. E. Billingham, eds. Pergamon Press, Macmillan, N. Y., pp. 144–164.

Ross, R., and T. K. Greenlee, Jr. 1966 Electron microscopy: attachment sites between connective tissue cells. Sci., 153: 997–999.

Sabatini, D. D., K. Bensch and R. J. Barrnett 1963 Cytochemistry and electron microscopy (The preservation of cellular ultrastructure and enzymatic activity by aldehdyde fixation). J. Cell Biol., 17: 19–58.

Salpeter, M. M. 1965 Disposition of nerve fibers in the regenerating limb of the adult newt, Triturus. J. Morph., 117: 201–211.

Salpeter, M. M., and M. Singer 1959 The fine structure of the adepidermal reticulum in the basal membrane of the skin of the newt, Triturus. J. Biophys. Biochem. Cytol., 6: 35–40.

——— 1960a Differentiation of the submicroscopic adepidermal membrane during limb regeneration in adult Triturus, including a note on the use of the term basement membrane. Anat. Rec., 136: 27–40.

——— 1960b The fine structure of mesenchymatous cells in the regenerating forelimb of the adult newt, Triturus. Develop. Biol., 2: 516–534.

——— 1962 The fine structure of mesenchymatous cells in the regenerating limbs of larval and adult Triturus. In: Electron Microscopy, vol. 2. S. S. Breese, Jr., ed. Academic Press, N. Y., pp. 00–12.

Schmidt, A. J. 1962 Distribution of polysaccharides in the regenerating forelimb of the adult newt, Diemictylus viridescens (Triturus v.). J. Exp. Zool., 149: 171–191.

——— 1966 The Molecular Basis of Regeneration: Enzymes. Illinois Monographs in Medical Sciences, 6(4), University of Illinois Press, Urbana.

Sheldon, H., and F. B. Kimball 1962 Studies on cartilage. III. The occurrence of collagen within vacuoles of the Golgi apparatus. J. Cell Biol., 12: 599–613.

Singer, M., and M. M. Salpeter 1961 Regeneration in vertebrates: the role of the wound epithelium. In: Growth in Living Systems. M. X. Zarrow, ed. Basic Books, N. Y., pp. 277–311.

Usuku, G., and J. Gross 1965 Morphologic studies of connective tissue resorption in the tail fin of metamorphosing bullfrog tadpole. Develop. Biol., 11: 352–370.

Watson, M. L. 1958 Staining of tissue sections for electron microscopy with heavy metals. J. Biophys. Biochem. Cytol., 4: 475–478.

Weiss, P. 1958 Cell contact. Int. Rev. Cytol., 7: 391–423.

Welsh, R. A. 1966 Intracytoplasmic collagen formations in desmoid fibromatosis. Am. J. Path., 49: 515–535.

PLATE 1

EXPLANATION OF FIGURES

1   One-day regenerate. Portion of a basal epithelial cell at the outer edge of the wound epithelium. The inner cell border is irregular and contains few hemidesmosomes (h). The adepidermal reticulum (ar) is disorganized. Granular endoplasmic reticulum (er); tonofilaments (t); vesicles (ve). × 15,750. Uranyl acetate and lead citrate.

2   Seven-day regenerate. Basal epithelial cells of apical wound epithelium showing a smooth inner cell border devoid of hemidesmosomes (small arrow) and disorganized glycocalyx (gl). The cell processes of a subapical macrophage (ma) are closely associated with those of an epithelial cell (large arrow). Surface folds (sf); tonofilaments (t); ingested red blood cell (rbc). × 6,132. Uranyl acetate and lead citrate.

PLATE 1

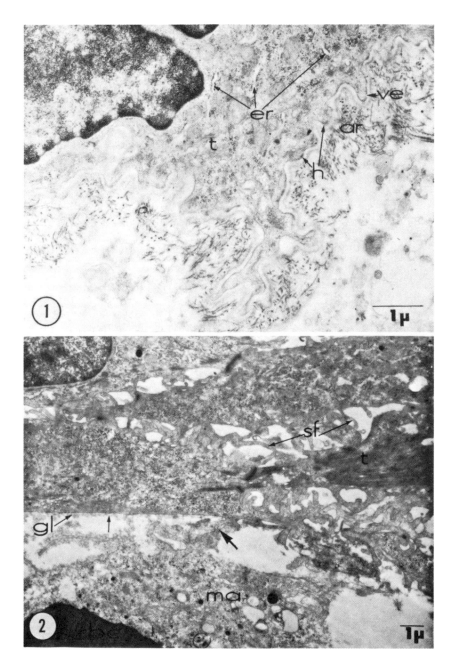

PLATE 2

EXPLANATION OF FIGURES

3 One-day regenerate. A wound epithelial cell containing cellular debris and melanosomes (me). Desmosomes (d). × 9,450. Uranyl acetate and lead citrate.

4 One-day regenerate. Peripheral nerve bundle near the amputated region of the forelimb showing axon vacuolization (large arrows) and myelin disorganization (small arrows). Two Schwann cells (s) are observed and surround normal-appearing nonmyelinated axons. × 32,130. Uranyl acetate and lead citrate.

PLATE 2

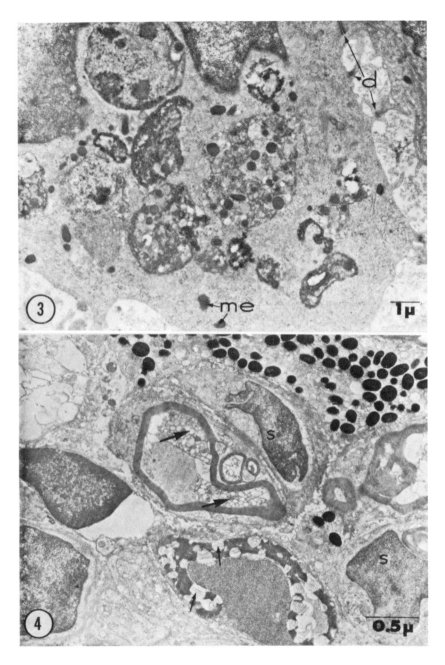

PLATE 3

5   One-day regenerate. Neural sheath (ns) of a peripheral nerve bundle showing disorganization and loss of the numerous invaginations as observed in normal peripheral nerve deep in stump tissues. Neurotubules (nt). $\times$ 18,480. Uranyl acetate and lead citrate.

6   Three-day regenerate. An isolated Schwann cell (s) enclosing normal appearing myelinated and nonmyelinated axons (na). $\times$ 4,200. Uranyl acetate and lead citrate.

PLATE 3

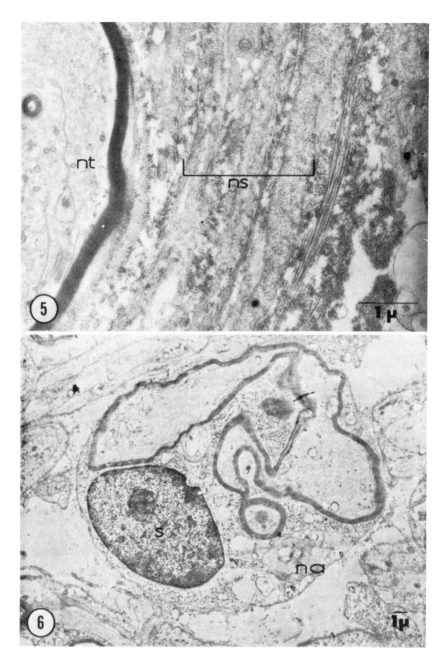

PLATE 4

EXPLANATION OF FIGURES

7   One-day regenerate. A portion of a degenerating skeletal muscle fiber
    containing a pyknotic nucleus (pn), a group of swollen mitochondria
    (m), bundles of myofilaments (mf), and vesicular portions of the
    sarcoplasmic reticulum (sr). × 8,400. Uranyl acetate and lead citrate.

8   Three-day regenerate. A portion of a normal-appearing skeletal mus-
    cle fiber containing a triad (tr) and a portion of a macrophage (ma)
    which appears to have penetrated the sarcolemma (large arrow). The
    cytoplasm of the macrophage contains membrane bound portions of
    myofilaments (md). × 17,325. Uranyl acetate and lead citrate.

144

PLATE 4

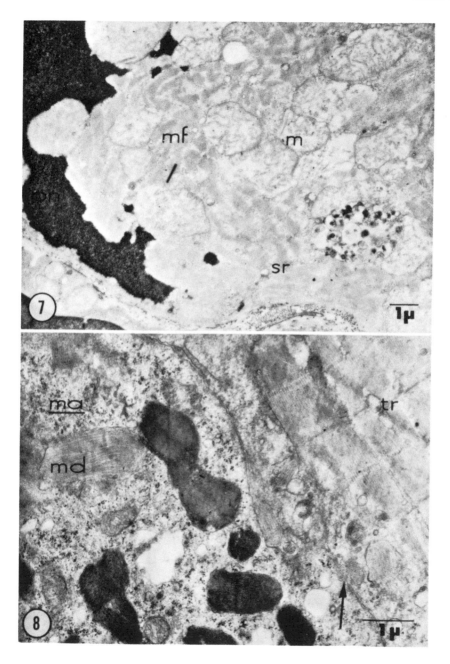

PLATE 5

9   One-day regenerate. Isolated portions of glycocalyx (large arrow) within a fibrogranular matrix. × 50,400. Uranyl acetate and lead citrate.

10 One-day regenerate. Portion of a thin myofiber showing scattered groups of myofilaments (mf) connected with electron-dense Z-bands (z). The cytoplasm contains many swollen vacuoles (sr) that may be remaining portions of the sarcoplasmic reticulum. The plasma membrane is studded with numerous invaginations (small arrows) and is adjacent to an irregularly folded glycocalyx (gl). Ribosomes on outer nuclear membrane (large arrows). × 8,400. Uranyl acetate and lead citrate.

PLATE 5

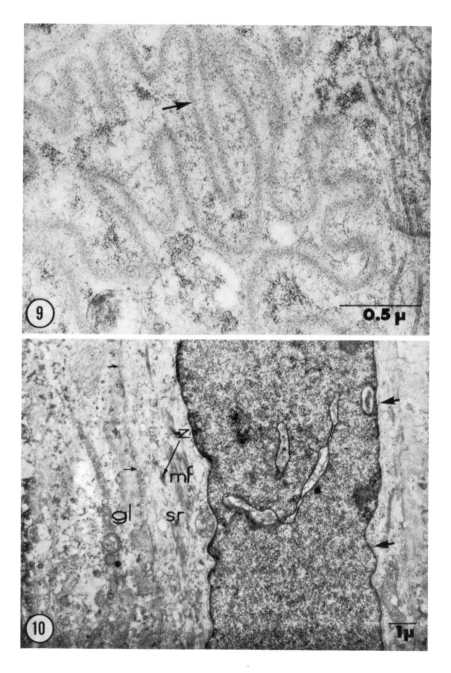

147

PLATE 6

EXPLANATION OF FIGURES

11 Three-day regenerate. Portions of five myofibers (1, 2, 3, 4, 5), one normal-appearing (5), the remaining four showing various degrees of reaction to amputation injury. Portions of two wandering cells (wc) are shown within the glycocalyx (gl) of the lower fibers. Myofibrils (mf). × 8,400. Uranyl acetate and lead citrate.

12 One-day regenerate. A portion of an injured myofiber showing numerous sarcoplasmic filaments (fil), isolated vacuoles of sarcoplasmic reticulum (sr), scattered polysomes (p) and mitochondria (m). Glycocalyx (gl). × 13,650. Uranyl acetate and lead citrate.

13 One-day regenerate. A portion of the sarcoplasm of a myofiber (adjacent to that of fig. 12) showing an intact triad (tr), disorganized myofilaments (mf) attached at electron-dense zones (z). Profiles of lamellae and vesicles are aggregated in a Golgi-like manner (g). × 23,520. Uranyl acetate and lead citrate.

148

PLATE 6

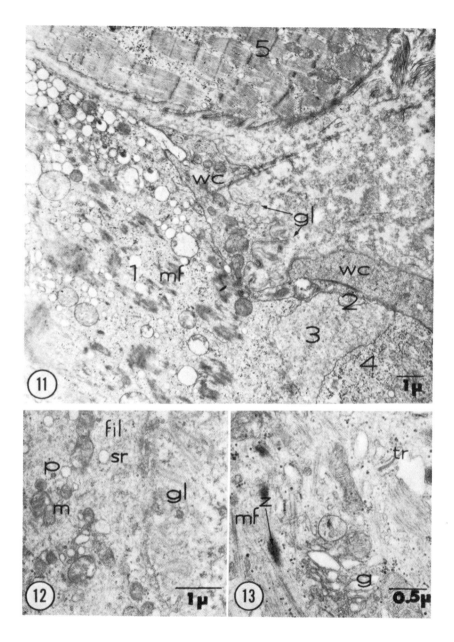

PLATE 7

EXPLANATION OF FIGURES

14 Seventeen-day regenerate. Wound epithelium showing a cuboidal
    basal cell (ep), melanocyte (me) and a polymorphonuclear leuco-
    cyte (pmn). The glycocalyx (gl) has reappeared. Two blastema cells
    (bl) are shown subapically. × 3,024. Uranyl acetate and lead citrate.

15 Twenty-two-day regenerate. A basal epithelial cell showing hemides-
    mosomes (h); a glycocalyx (gl); and an early organizing adepidermal
    reticulum (ar). Blastema cell (bl). × 8,904. Uranyl acetate and lead
    citrate.

PLATE 7

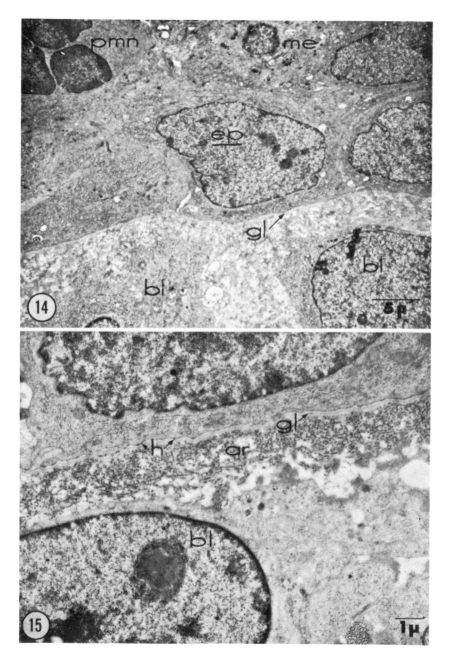

16 Seventeen-day regenerate. Portions of three blastema cells (1, 2, 3) separated by an intercellular space with few collagen fibrils (arrow). One cell shows a large Golgi zone (g) and granular endoplasmic reticulum (er). Nerve (ne). × 5,678. Uranyl acetate and lead citrate.

17 Seventeen-day regenerate. Portions of three blastema cells (1, 2, 3). Cytoplasmic processes contain free ribosomes (r). Granular endoplasmic reticulum (er); collagen (co). × 5,678. Uranyl acetate and lead citrate.

PLATE 8

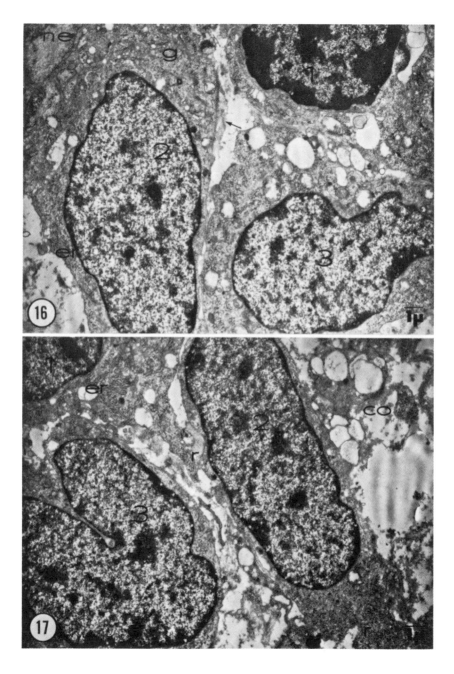

PLATE 9

18 Twenty-two-day regenerate. Portions of two closely applied blastema cells (small arrows) surrounded by an irregularly placed glycocalyx (large arrows). On cell contains profiles of granular endoplasmic reticulum (er) and polysomes (p). × 8,257. Uranyl acetate and lead citrate.

19 Twenty-two-day regenerate. Portions of blastema cells with a more regularly placed glycocalyx (gl). Collagen fibrils are sparse (arrows). × 8,257. Uranyl acetate and lead citrate.

PLATE 9

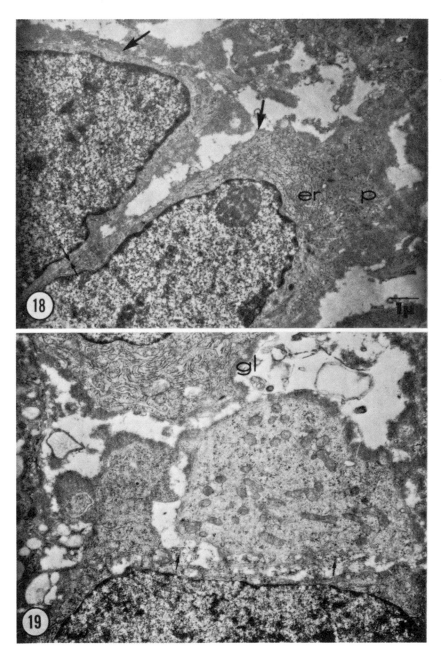

20 Seventeen-day regenerate. Adhesion discs (arrows) between two blastema cells. × 100,800. Uranyl acetate and lead citrate.

21 Twenty-three-day regenerate. Adhesion disc (arrow) between two blastema cells. × 41,126. Uranyl acetate and lead citrate.

22 Seventeen-day regenerate. Sarcoplasmic extensions of a skeletal muscle fiber. Sarcolemmal thickenings (large arrow) occur along the extensions which are adjacent to collagen fibrils (co). Polysomes (p) are shown in some of the extensions. Direction of blastema (bl); myofilaments (mf). × 17,640. Uranyl acetate and lead citrate.

PLATE 10

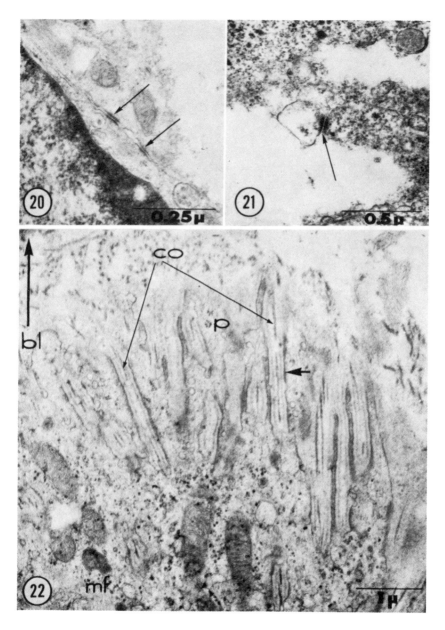

23 Seventeen-day regenerate. A typical fibroblast (fi) located within the endomysium of skeletal muscle fibers along the deep border of the blastema. Direction of blastema (bl); lipid (li); myofibrils (mf); z-bands (z). × 8,736. Uranyl acetate and lead citrate.

PLATE 11

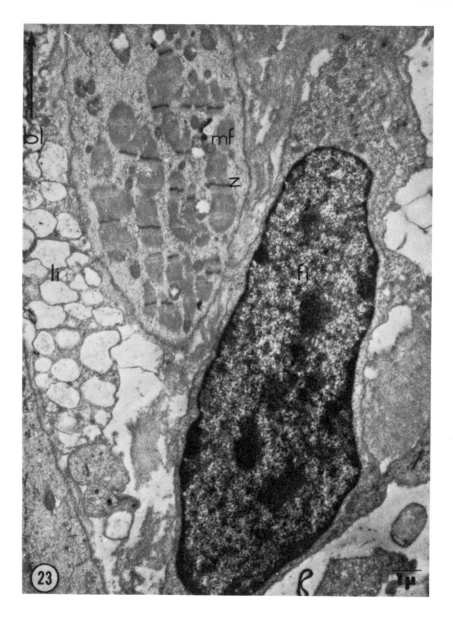

PLATE 12

EXPLANATION OF FIGURE

24 Seventeen-day regenerate. An endomysial fibroblast showing intracellular banded collagen-like fibrils (small arrows) and extracellular
banded fibrils in close proximity to intracellular filaments (large
arrow). A condensation of cytoplasmic filaments (fil) are shown beneath the plasma membrane. Direction of blastema (bl). × 8,736.
Uranyl acetate and lead citrate.

PLATE 12

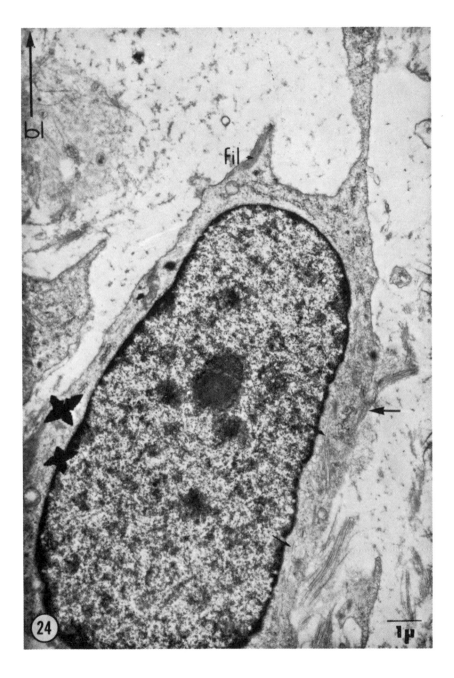

PLATE 13

EXPLANATION OF FIGURE

25 Seventeen-day regenerate. A portion of the cytoplasm of a fibroblast
containing groups of banded fibrils, partly enclosed by smooth membranes (arrows). Filaments (fil) are shown in one of the cytoplasmic
extensions. A portion of an eosinophile (e) is also seen in the field.
$\times$ 12,879. Uranyl acetate and lead citrate.

PLATE 13

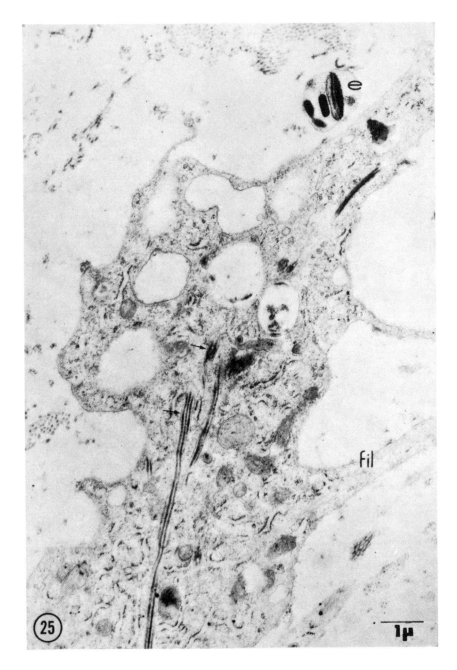

PLATE 14

26 Seventeen-day regenerate. Cross-section of a portion of a fibroblast showing groups of fibrils cut both transversely (small arrows) and longitudinally. The membranes surrounding some fibrils appear to be continuous with the plasma membrane (large arrows). × 22,050. Uranyl acetate and lead citrate.

27 Seventeen-day regenerate. A portion of a fibroblast showing banded fibrils near the nucleus as well as near the plasma membrane (arrows). × 17,615. Uranyl acetate and lead citrate.

PLATE 14

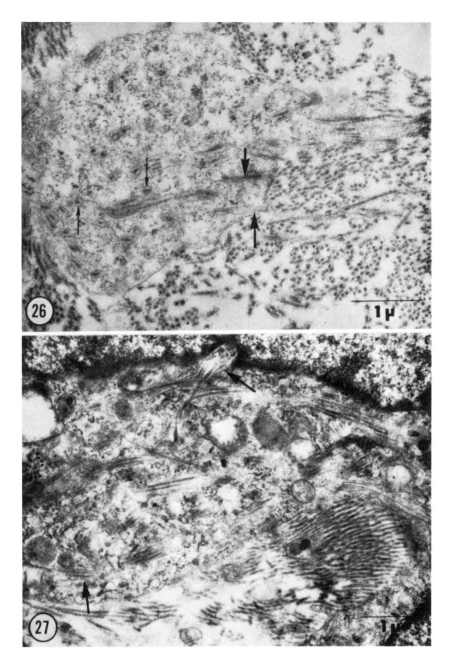

PLATE 15

28 Seventeen-day regenerate. A portion of the cytoplasm of a fibroblast showing profiles of Golgi lamellae and vesicles (g). Membrane-bounded fibrils are found near the Golgi (1) and also near the plasma membrane (2). Mitochondria (m). $\times$ 44,410. Uranyl acetate and lead citrate.

166

PLATE 15

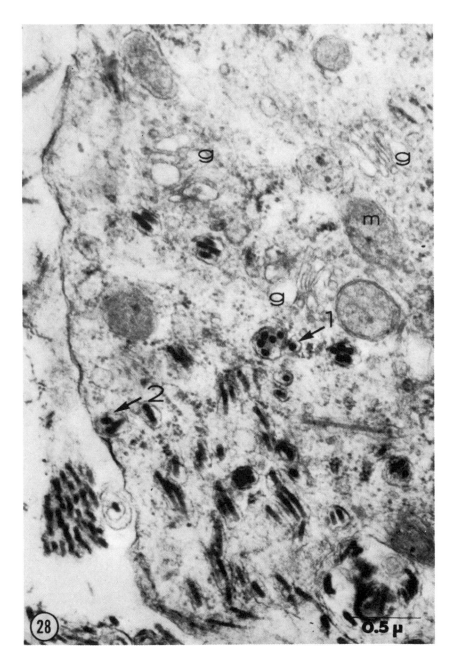

29   Fifty-eight-day regenerate. A perichondrial cell (pc) and a chondroblast (ch) of developing skeleton. An electron-dense precipitate is
seen around the chondroblast within a collagenous matrix (co).
× 13,650. Uranyl acetate and lead citrate.

PLATE 16

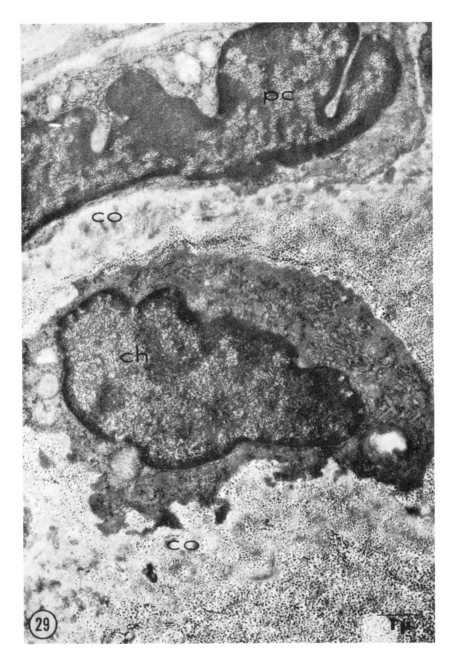

Histochemical Identification of Sulfated and Non-sulfated Mucopolysaccharides in Regenerating Forelimbs of Adult Urodeles

D. J. PROCACCINI and CATHERINE M. DOYLE

Introduction

The urodele forelimb is known to contain sulfated and non-sulfated mucopolysaccharides and glycogen. Marked quantitative and qualitative alterations in these metabolites occur as a result of forelimb amputation and subsequent regeneration [1, 7, 8, 9]. During wound healing and redifferentiation polysaccharide reserves are depleted and mucopolysaccharide components of connective tissue, bone, and cartilage are broken down. Dedifferentiation of blastema cells results in *de novo* synthesis of these substances. Prochondral cells actively engage in the production of a sulfated mucopolysaccharide for incorporation into the cartilagenous matrix [1, 7]. Non-sulfated mucopolysaccharides are prevalent in developing connective tissues, dermal gland secretions and the epidermal-dermal basement membrane; polysaccharide glycogen is present in late blastema cells, developing muscle, cartilage and the apical cap [8].

We have developed a rapid histochemical procedure which simultaneously demonstrates these tissue carbohydrates in both intact and regenerating urodele forelimbs. The process, a combination of the alcian blue [5] and periodic acid-Schiff techniques, is carried out on formalin fixed, versene decalcified tissue. Alcian blue is believed to form salt linkages with the anionic SO_3H of sulfated mucopolysaccharides [6]. Periodic acid-Schiff characteristically imparts a red-purple phenyl methane dye to glycogen and neutral glycosaminoglycans [4]. This simultaneous demonstration of both categories of polysaccharides offers obvious advantages which we have utilized in investigating histogenetic events in regenerating forelimbs of the newt *Diemictylus viridescens*. The results of this study are reported here.

Materials and Methods

Regenerating *Diemicytlus viridescens* forelimbs, at different stages, were fixed in buffered formalin, decalcified 48 h in two changes of 5.5% versene, dehydrated, cleared in xylene and embedded in paraffin. Ten micron serial sections were mounted on albuminized slides. Ethyl alcohol (90%) was used as a floating medium to prevent excess glycogen dissolution. Hydrated slides were rinsed three minutes in 3% acetic acid, stained two hours in alcian blue (pH 2.6), rinsed again (3 min) in 3% acetic acid, then placed in distilled water (3 min). Slides were next treated 5 min with cold 0.5% periodic acid, rinsed, treated with cold Schiff's reagent 15 min, transferred through two changes of cold sodium sulfite (5 cc 10% Na metabisulfite, 5 cc 1 N HCl, 90 cc H_2O), rinsed in cold water 5 min, stained with Delafield's hematoxylin [2] 30 sec, dehydrated, cleared and mounted. To demonstrate glycogen, control slides were digested with buffered 1% diastase of malt (50 cc Sorensen's Buffer, pH 6.8, plus 0.5 g diastase of malt), at 20° C for 15 min, then rinsed prior to periodic acid treatment.

Results

Tissue periodic acid-Schiff reactivity obtained in the present investigation augments and extends Schmidt's [8] earlier observations. Non-sulfated mucopolysaccharide response was noted particularly in outer keratinized epithelial cells of normal and apical cap epithelium, the adepidermal reticulum, the secretions of dermal glands, sarco-

---

Fig. 1. D. viridescens forelimb 1 week after amputation. × 64.
Fig. 2. D. viridescens forelimb 4 weeks after amputation. × 64.
Fig. 3. Detail, fig. 2, developing cartilage. × 156. (a) apical cap. Outer keratinized epithelial layer, strong positive PAS; (rc) regressing cartilage, metachromatic alcian blue—PAS positive. Mature cartilage, strong positive alcian blue; (dc) developing cartilage, intensity of alcian blue increases from weak—moderate in distal-proximal direction; (b) bone, positive PAS; (m) muscle, positive PAS. Arrow indicates areas of glycogen concentration.

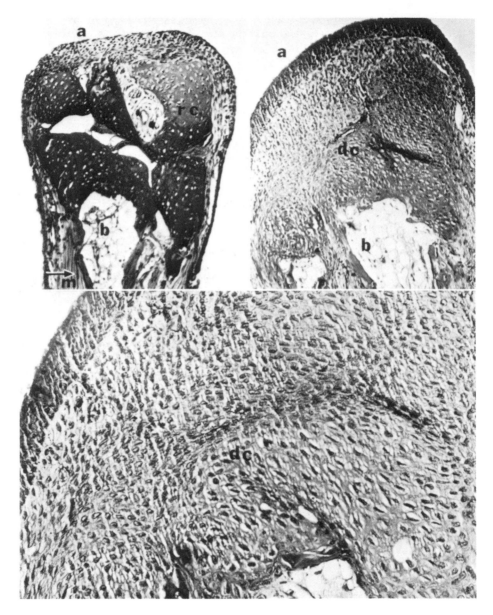

Fig. 1–3

plasm, erythrocytes, and bone matrix; dedifferentiating muscle and connective tissue and blastema cells were weakly responsive. Although glycogen stores are somewhat depleted due to decalcification and alcian blue staining, glycogen was apparent in late blastema cells, mature and to a lesser extent in maturing muscle, and wound epithelium. Alcian blue response was strong and regular in mature cartilage extracellular matrix. Connective tissue fibrils, located intermuscularly and dermally, as well as the cytoplasm of the basal layers of the wound epithelium also took up the alcian blue stain. The intense alcian blue response of mature cartilage is decreased during regressive phases in areas of osteoclastic activity. Both PAS and alcian blue responses are apparent in dedifferentiating cartilage. Fine alcian blue responsive fibrils became discernible in the blastema prior to the morphologically demonstrable cone stage. Extracellular fibrillar alcian blue response parallels the proximo-distal direction of cartilage formation known to occur in regenerating appendages [3].

Discussion

The distribution of PAS responsive polysaccharide glycogen and non-sulfated mucopolysaccharides in regenerating appendages has been described previously [9]. Alcian blue responses have not been extensively investigated in regenerating urodele tissue. The present procedure demonstrates a technique by which these polysaccharides may be seen in the same section, and enables the investigator to study simultaneously the changes which occur in these two categories of polysaccharides during regeneration. This technique is particularly valuable in the study of the biochemistry of antimetabolite-induced inhibition of regeneration which we are currently investigating[1].

References

1. Anton, H. J.: The origin of blastema cells and protein synthesis during forelimb regeneration in Triturus; in V. Kiortsis and H. A. L. Trampusch Regeneration in animals and related problems, p. 377 (North Holland Publishing Co., Amsterdam 1965).
2. Carlton, H. M. and Leach, F. H.: Histological technique (Oxford University Press, New York 1947).

[1] Procaccini, D. J. and C. M. Doyle (in preparation).

3. HAY, E. D.: Metabolic patterns in limb development and regeneration; in R. K. DEHAAN and H. URSPRUNG Organogenesis, p. 315 (Holt Rinehart and Winston, New York 1965).
4. JEANLOZ, R. W.: Mucopolysaccharides (acidic glycosaminoglycans); in M. FLORKIN and E. H. STOTZ Comprehensive biochemistry, vol. 5, p. 296 (Elsevier, New York 1963).
5. MOWRY, R. W.: The special value of methods that color both acidic and vicinal hydroxyl groups in the histochemical study of mucins, with revised directions for the colloidal iron stain, the use of alcian blue G 8X and their combinations with the periodic acid-Schiff reaction. Ann. N. Y. Acad. Sci. *106:* 402 (1963).
6. PEARSE, A. G. E.: Histochemistry: Theoretical and applied, 2nd ed., p. 261 (Little, Brown and Co., Boston 1961).
7. SAITO, T.: Autoradiographic study on the ^{35}S sulfate metabolism in the regenerating tissues of the adult newt forelimb. Embryologia *9:* 279 (1967).
8. SCHMIDT, A. J.: Distribution of polysaccharides in the regenerating forelimb of the adult newt, *Diemicytlus viridescens (Triturus v.).* J. exp. Zool. *149:* 171 (1962).
9. SCHMIDT, A. J.: The cellular biology of vertebrate regeneration and repair, p. 131 (The University of Chicago Press, Chicago 1968).

174

AUTHOR INDEX

KEY-WORD TITLE INDEX